Careers
in Focus

Manufacturing

Ferguson Publishing Company
Chicago, Illinois

670
CAR

Copyright © 2000 Ferguson Publishing Company
ISBN 0-89434-314-9

Library of Congress Cataloging-in-Publication Data

Careers in focus. Manufacturing.
        p. cm.
     ISBN 0-89434-314-9
     1. Industrial technicians-Vocational guidance. 2. Industrial engineer
ing-Vocational guidance. I. Title: Manufacturing.

TA 158.C36 2000
670'.23-dc21

                                                      00-022829

Printed in the United States of America

Cover photo courtesy Tony Stone Images

Published and distributed by
Ferguson Publishing Company
200 West Jackson Boulevard, 7th Floor
Chicago, Illinois 60606
800-306-9941
www.fergpubco.com

X-3

Table of Contents

Introduction

Manufacturing covers a wide range of industries, including food, beverage, military operations, iron and steel, textiles, lumber, tobacco, automobile, aerospace, and petrochemical. In manufacturing, there are two types of goods produced: durable and nondurable. Durable goods have a long life span and hold up over time; examples of durable goods are cars, airplanes, and washing machines. Nondurable goods have a shorter life span and include such products as food, cosmetics, and clothing.

No matter what industry, certain key structural components exist in different types of manufacturing operations. There are many kinds of engineers whose work focuses on research, development, analysis, planning, survey, application, facility evaluation, and more. Sometimes engineers manage a staff or manage projects from start to finish. Engineers whose work entails research and development refine the production process and make recommendations to their companies based on their research findings. Engineers work with computers, in robotics development and implementation, and plant safety. Technicians, who work closely with and assist engineers, help to execute various projects and to conduct research by running tests and fulfilling other duties. Production workers comprise the largest group of manufacturing employees. They operate and maintain the computers, machines, and components that turn raw materials into finished products.

Today there are a declining number of manufacturing jobs, and those jobs that exist are less viable than they were 10 or 20 years ago for various reasons. One reason is that many factories have been and continue to be relocating to foreign countries, taking jobs with them. As a result, labor unions have lost some of their strength to negotiate for better contracts and wages for manufacturing workers. The other key reason for the continuing decrease in factory jobs is that it is an economic advantage for companies to reduce their labor force by replacing it with computer technologies and automated, robotics-based machinery. In many cases, these machines are more efficient and productive than human workers. And to remain competitive, many companies work to become even more automated, which will in turn eliminate more jobs. However, while many assembly line jobs will disappear, the engineering and technical jobs for those who develop and refine the technologies behind the automated machinery should increase over the next 10 years.

Each article in this book discusses a particular manufacturing occupation in detail. The information comes from Ferguson's *Encyclopedia of Careers and Vocational Guidance*. The History section describes the history of the particular job as it relates to the overall development of its industry or field. The Job describes the primary and secondary duties of the job. Requirements discusses high school and postsecondary education and training requirements, any certification or licensing necessary, and any other personal requirements for success in the job. Exploring offers suggestions on how to gain some experience in or knowledge of the particular job before making a firm educational and financial commitment. The focus is on what can be done while still in high school (or in the early years of college) to gain a better understanding of the job. The Employers section gives an overview of typical places of employment for the job. Starting Out discusses the best ways to land that first job, be it through the college placement office, newspaper ads, or personal contact. The Advancement section describes what kind of career path to expect from the job and how to get there. Earnings lists salary ranges and describes the typical fringe benefits. The Work Environment section describes the typical surroundings and conditions of employment—whether indoors or outdoors, noisy or quiet, social or independent, and so on. Also discussed are typical hours worked, any seasonal fluctuations, and the stresses and strains of the job. The Outlook section summarizes the job in terms of the general economy and industry projections. For the most part, Outlook information is obtained from the Bureau of Labor Statistics and is supplemented by information taken from professional associations. Job growth terms follow those used in the *Occupational Outlook Handbook:* Growth described as "much faster than the average" means an increase of 36 percent or more. Growth described as "faster than the average" means an increase of 21 to 35 percent. Growth described as "about as fast as the average" means an increase of 10 to 20 percent. Growth described as "little change or more slowly than the average" means an increase of 0 to 9 percent. "Decline" means a decrease of 1 percent or more.

Each article ends with For More Information, which lists organizations that can provide career information on training, education, internships, scholarships, and job placement.

Automotive Industry Workers

School Subjects

Mathematics
Technical/Shop

Personal Skills

Following instructions
Mechanical/manipulative

Work Environment

Primarily indoors
Primarily one location

Minimum Education Level

High school diploma

Salary Range

$20,000 to $40,000 to $100,000

Certification or Licensing

Voluntary

Outlook

Little change or more slowly than
the average

Overview

Automotive industry workers are the people who work in the parts production and assembly plants of automobile manufacturers. Their labor produces automobiles from the smallest part to the completed automobiles. Automotive industry workers read specifications; design parts; build, maintain and operate machinery and tools used to produce parts; inspect parts and assemble the automobiles.

History

In our mobile society, it is difficult to imagine a time without automobiles. Yet just over 100 years ago, there were none. In the late 1800s, inventors were just beginning to tinker with the idea of a self-propelled vehicle. Early experiments used steam to power a vehicle's engine. Two German engineers developed the first internal combustion engine fueled by gasoline. Karl Benz finished the first model in 1885 and Gottlieb Daimler finished building a similar model in 1886. Others around the world had similar successes in the late 1800s and early 1900s. In these early days, no one imagined people would become so reliant on the automobile as a way of life. In 1898, there were 50 automobile manufacturing companies in the United States, a number that rose to 241 by 1908.

Early automobiles were expensive to make and keep in working order and could only be used to travel short distances; they were "toys" for men who had the time and money to tinker with them. One such man was Henry Ford. He differed from others who had succeeded in building automobiles in that he believed the automobile could appeal to the general public if the cost of producing automobiles was reduced. The Model A was first produced by the Ford Motor Co. in small quantities in 1903. Ford made improvements to the Model A, and in October 1908, found success with the more practical Model T. The Model T was the vehicle that changed Ford's fortune and would eventually change the world. It was a powerful car with a possible speed of 45 miles per hour that could run 13 to 21 miles on a gallon of gasoline. Such improvements were made possible by the use of vanadium steel—a lighter and more durable steel. Automobiles were beginning to draw interest from the general public as newspapers reported early successes, but they were still out of reach for most Americans. The automobile remained a curiosity to be read about in the newspapers until 1913. That's when Ford changed the way his workers produced automobiles in the factory. Before 1913, skilled craftsmen made automobiles in Ford's factory, but Ford's moving assembly line reduced the skill level needed and sped up production. The moving assembly line improved the speed of chassis assembly from 12 hours and 8 minutes to 1 hour and 33 minutes. Craftsmen were no longer needed to make the parts and assemble the automobiles. Anyone could be trained for most of the jobs required to build an automobile in one of Ford's factories, making it possible to hire unskilled workers at lower wages.

For many early automotive workers, Ford's mass production concept proved to be both a blessing and a curse. Demand was growing for the affordable automobile, even during the Depression years, bringing new jobs for people who desperately needed them. However, working on an assembly line could be tedious and stressful at the same time. Ford paid his workers well (he introduced the five-dollar day in 1914, a high wage for the time) but he demanded a lot of them. He sped up the assembly line on several occasions, and many workers performed the same task for hours at a frenzied pace, often without a break.

Such conditions led workers to organize unions and, through the years, workers have gained more control over the speed at which they work and pay rates. Many of today's automotive industry workers belong to unions, such as the United Auto Workers (founded in 1936). The industry continued to evolve with automotive technology in the 1940s and 1950s. American automobiles were generally large and consumed a lot of gasoline, but a strong U.S. economy afforded many Americans the ability to buy and maintain such vehicles. In Europe and Japan, smaller, fuel-efficient cars were more popular. This allowed foreign automakers to cut deeply into the American automobile market during fuel shortages in the 1970s. Automotive workers suffered job cuts in the 1980s because of declining exports and domestic sales. Today, the industry has recovered from the losses of the 1980s largely by producing vehicles that can compete with fuel-efficient, foreign made ones. Also, trade agreements have encouraged foreign automakers to build manufacturing plants in the United States, creating new jobs for U.S. workers. In 1996, the United States led the world in vehicle production (12 million vehicles), but only with the contribution of foreign automakers with plants here. Without such involvement, the United States would be second in worldwide vehicle production to Japan, according to the Association of International Automobile Manufacturers.

The Job

The term automotive production worker covers the wide range of people who build the 12 million cars produced in the United States each year. Automotive industry workers are employed in two types of plants: parts production plants and assembly plants. Similar jobs are also found with companies that manufacture farm and earth-moving equipment and their workers often belong to the same unions and undergo the same training. Major automobile manufacturers are generally organized so that automobiles are assembled at a few large plants that employ several thousand workers. Parts for the automobiles are made at many smaller plants which may employ fewer than 100 workers. Some plants that produce parts are not owned by the automobile manufacturer, but may in fact be an independent company that specializes in making one important part. These independent manufacturers may supply parts to several different automobile makers.

Whether they work in a parts plant or an assembly plant, automotive workers are generally people who work with their hands, spend a lot of time standing, bending, lifting, and doing a lot of repetitive work. They often work in noisy areas and are required to wear protective equipment throughout their work day, such as safety glasses, earplugs, gloves, and masks. Because automotive industry workers often work in large plants that operate 24 hours a day, they usually work in shifts. Shift assignments are generally made on the basis of seniority.

Some of the common jobs for automotive industry workers are:

Precision metalworkers: This is one of the more highly skilled positions found in automotive production plants. Precision metalworkers create the metal tools, dies, and special guiding and holding devices that produce automotive parts—thus, they are sometimes called tool and die makers. They must be familiar with the entire manufacturing process and have knowledge of mathematics, blueprint reading, and the properties of metals, such as hardness and heat tolerance. Precision metalworkers may perform all or some of the steps needed to make machining tools, including reading blueprints, planning the sequence of operations, marking the materials, and cutting and assembling the materials into a tool. Precision metalworkers often work in quieter parts of the production plants.

Machinists and *tool programmers:* Machinists make the precision metal parts needed for automobiles using tools such as lathes, drill presses, and milling machines. In automotive production plants, their work is repetitive as they generally produce large quantities of one part. Machinists may spend their entire shift machining the part. Some machinists also read blueprints or written specifications for a part. They may also calculate where to cut into the metal, how fast to feed the metal into the machine, or how much of the metal to remove. Machinists may also select tools and materials needed for the job and mark the metal stock for the cuts to be made. Increasingly, the machine tools used to make automotive parts are computerized. Computer numerically controlled machining is widespread in many manufacturing processes today. This is where tool programmers come in. Tool programmers write computer programs that direct the machine's operations and machinists monitor the computer-controlled process.

Maintenance workers: This is a vague term often used among industry workers. People who work in maintenance may do a number of things. They may repair or make new parts for existing machines. They also may set up new machines. A maintenance worker may work with sales representatives from the company that sold the automobile manufacturer the piece of equipment. Maintenance workers are responsible for the upkeep of such machines and should be able to perform all of the machine's operations.

Welders: Welders use equipment that joins metal parts by melting and fusing them to form a permanent bond. There are different types of welds and other equipment to make the welds. In manual welding, the work is entirely controlled by the welder. Other work is semi-automatic, in which machinery such as a wire feeder is used to help perform the weld. Much of the welding work in automotive plants is repetitive; in some of these cases welding machine operators monitor machines as they perform the welding tasks. Because they work with fire, welders must wear a lot of safety gear such as protective clothing, safety shoes, goggles, and hoods with protective lenses.

Inspectors: Inspectors check the manufacturing process at all stages to ensure products meet quality standards. Everything from raw materials to parts to the finished automobile is checked for dimension, color, weight, texture, strength, or other physical characteristics, as well as proper operation. Inspectors identify and record any quality problems and may work with any of several departments to remedy the flaw. Jobs for inspectors are declining because at many stages, inspection has become automated. Also, there is a move to have workers self-check their work on the production line.

Supervisors: Management structure and philosophy differs greatly by automobile manufacturer. In many cases, floor or line supervisors are responsible for a group of workers who produce one part or perform one step in a process. Floor or line supervisors may report to department heads or foremen who oversee several such departments. Many supervisors are production workers who have worked their way up the ranks; still others have a management background and in many cases, a college degree in business or management.

Requirements

High School

Many automotive production jobs require mechanical skills, so students should take advantage of any shop programs their high school offers, such as auto mechanics, electronics, welding, drafting, and computer programming and design. In the core subject areas, mathematics including algebra and geometry is useful for reading blueprints and computer programs that direct machine functions. Chemistry is useful for the many workers who will need to be familiar with the properties of metals. English classes are also important to help someone who wants to work in a factory setting communicate both verbally with supervisors and coworkers and be able to read and understand complex instructions.

Postsecondary Training

Many of the jobs in an automotive plant are classified as semiskilled or unskilled positions, and people with some mechanical aptitude, physical ability, and a high school diploma are qualified to do them. However, there is often stiff competition for jobs with large automakers like GM, Ford, and Chrysler because they offer good benefits and pay compared to jobs that require similar skill levels. Therefore, those with some postsecondary training, certification, or experience stand a better chance at getting a job in the automotive industry than someone with only a high school diploma.

Formal training for machining, welding, and toolmaking is offered in vocational schools, vocational-technical institutes, community colleges, and private schools. Increasingly, such postsecondary training or certification is the route many workers take to getting an automotive industry job. In the past, apprenticeships and on-the-job training were the routes many workers took to get factory jobs, but these options are not as widely available today. Electricians, who generally must complete an apprenticeship, may find work in automotive plants as maintenance workers.

Certification or Licensing

Certification is available but not required for many of the positions in an automotive production plant. The American Welding Society supports education and training for welders. For precision metalworkers and machinists, the National Tooling and Machining Association operates training centers and apprentice programs and sets skill standards.

Other Requirements

Working in an automotive production plant can be physically challenging. For many jobs, workers need the physical capability to stand for long periods, lift heavy objects, and maneuver hand tools and machinery. Of course, some jobs in an automotive production plant can be performed by a person with a physical disability. For example, a person who uses a wheelchair may work well on an assembly line job that requires only the use of his or her hands. Automotive workers should have hand and finger dexterity and the ability to do repetitive work accurately and safely.

Exploring

Students interested in a career in the automotive industry should explore the diverse opportunities the field offers. Consider your current talents and interests in making such a decision. Do you enjoy working with your hands? Following complex instructions? Do you think you could do repetitive work on a daily basis? Are you a natural leader who would enjoy a supervisory position? Once you have an idea what area you want to pursue, the best way to learn more is to find someone who does the job and think of questions to ask them. Assembly plants are generally located in or near large cities, but if you live in a rural area you can still probably find someone with a similar job at a parts plant or other manufacturer. Even small towns generally have machine shops or other types of manufacturing plants that employ machinists, tool and die makers, inspectors, and other production workers. Local machine shops or factories are a good place to get experience, perhaps through a summer or after-school job to see if you enjoy working in a production environment. Many high schools have cooperative programs that employ students who want to gain work experience in their area of interest.

Employers

Automotive production workers can find jobs with both domestic automakers, such as the Big Three, and with foreign automakers like Mitsubishi and Honda, which both have large assembly plants in the United States. Large assembly plants may employ several thousand workers. Parts production plants may employ fewer workers but there are more of these plants. Assembly plants are generally located in or near large cities, especially in the Northeast and Midwest where heavy manufacturing is concentrated. Parts production plants vary in size, from a few dozen workers to several hundred. Employees of these plants may all work on one small part or on several parts that make up one component of an automobile. Parts production plants are located in smaller towns as well as urban areas. The production processes in agricultural and earth-moving equipment factories are similar to those in the automotive industry, and workers trained in welding, toolmaking, machining, and maintenance may find jobs with companies like Caterpillar and John Deere.

Starting Out

Hiring practices at large plants are usually very structured. Such large employers generally don't place "help wanted" ads. Rather, they accept applications year-round and keep them on file. Applicants generally complete an initial application and may be placed on a hiring list. Others get started by working as temporary or part-time workers at the plant and using their experience and contacts to obtain full-time, permanent positions. Some plants work with career placement offices of vocational schools and technical associations to find qualified workers. Others may recruit workers at job fairs. Also, as with many large factories, people who have a relative or know someone who works at the plant usually have a better chance of getting hired. Their contact may put in a good word with a supervisor or advise them when an opening occurs.

Advancement

Automotive production plants are very structured in their paths of advancement. Large human resources departments oversee the personnel structures of all departments; each job has a specific description with specific qualifications. Longevity is usually the key to advancement in an automotive plant. For many, advancement means staying in the same position and moving up on the salary scale. Others acquire experience and, often, further training, to advance to a position with a higher skill level, more responsibility, and higher pay. For example, machinists may learn a lot about many different machines throughout their careers and may undergo training or be promoted to become precision metalworkers. Others with years of experience become supervisors of their departments.

Earnings

Salaries vary widely for automotive production workers, depending on what their job is and how long they've been with the company. Supervisors may earn $40,000 to $50,000 a year or more, depending on the number of people they supervise. Pay for semiskilled or unskilled workers, such as assemblers, is considerably lower, usually in the $20,000 range. Still, such production jobs are sought after because this pay is higher than such workers may find elsewhere based on their skill level. Median annual earnings for welders in 1996 was $24,856; median earnings for machinists in 1996 was $28,600. Earnings are usually much higher for workers who are members of a union and employed by a Big Three automaker. Few of these workers earn less than $40,000 a year, and some earn as much as $100,000 a year because of mandatory overtime and six- or seven-day work weeks.

Workers employed by large, unionized companies such as Ford and Chrysler enjoy good benefits, including paid health insurance, paid holidays, sick days, and personal days. Large employers generally offer retirement plans and many match workers' contributions to retirement funds. Automotive production workers who work for independent parts manufacturers may not enjoy the comprehensive benefit programs that employees of large companies do, but generally are offered health insurance and paid personal days.

Work Environment

Working as a production worker in an automotive plant can be stressful, depending on the worker's personality, job duties, and management expectations. Assembly line workers have little control over the speed at which they must complete their work. They can generally take breaks only when scheduled. Norm Ritchie, a machine operator at a Chrysler parts plant in Perrysburg, Ohio, said the job can be stressful: "The pressure [of the assembly line] affects people in different ways. Sometimes people get pretty stressed out; other people can handle it." Ritchie, who works on steering shafts, also said that noise is a concern in his area of the plant. He estimated that the noise level is about 90 decibels all the time. Automotive production workers must follow several safety precautions every day, including wearing protective gear (such as earplugs) and undergoing safety training throughout their careers.

Outlook

Job growth is not expected for the U.S. automotive industry as the new century begins. The industry has recovered from the 1980s, which saw steep layoffs and job losses. Automotive production employment is expected to remain steady at best. The industry reached its peak employment level in 1979 with 1.1 million workers. But fuel shortages in the late 1970s and early 1980s made the larger American-made cars less appealing than foreign cars, which for years had been made smaller and more fuel efficient. By 1982, the industry employed only 600,000 workers. In the 1990s, employment has hovered around 700,000. As of fall 1998—the most recent figure available—756,700 workers were employed in the industry, according to the federal Bureau of Labor Statistics. Today, American automakers operate leaner manufacturing facilities, much like their Japanese counterparts, striving for higher efficiency with fewer workers. To hold down personnel costs, manufacturers often choose to increase the individual worker's responsibility and offer more overtime rather than hiring new workers. However, the decline in employment among American-owned automakers has been balanced by new foreign-owned manufacturing plants that have been built in the United States. Today, many U.S. automotive workers are employed by foreign-owned automakers such as Honda and Mitsubishi. Also, a strong domestic economy in the late 1990s has made it easier for consumers to purchase automobiles, keeping production demands steady.

For More Information

These professional societies promote the skills of their trades and can provide career information.

National Tooling & Machining Association
9300 Livingston Road
Fort Washington, MD 20744
Tel: 800-248-NTMA
Web: http://www.ntma.org

American Welding Society
550 NW Lejeune Road
Miami, FL 33126-5699
Tel: 800-443-9353
Web: http://www.aws.org

These are two of many unions that represent automotive production workers. They can provide information about training and education programs in your area.

United Auto Workers
8000 E. Jefferson
Detroit, MI 48214
Tel: 313-926-5000
Web: http://www.uaw.org

International Association of Machinists and Aerospace Workers
9000 Machinists Place
Upper Marlboro, MD 20772-2687
Tel: 301-967-4500
Web: http://www.iamaw.org

Dairy Products Manufacturing Workers

	School Subjects
Agriculture	
Biology	
Chemistry	
	Personal Skills
Following instructions	
Technical/scientific	
	Work Environment
Primarily indoors	
Primarily one location	
	Minimum Education Level
High school diploma	
	Salary Range
$15,000 to $27,500 to $40,000+	
	Certification or Licensing
Required by all states	
	Outlook
Decline	

Overview

Dairy products manufacturing workers set up, operate, and tend continuous-flow or vat-type equipment to process milk, cream, butter, cheese, ice cream, and other dairy products following specified methods and formulas.

History

Since herd animals were first domesticated, humankind has kept cattle for meat and milk. From its ancient beginnings in Asia, the practice of keeping cattle spread across much of the world. Often farmers kept a few cows to supply their family's dairy needs. Because fresh milk spoils easily, milk that was not consumed as a beverage had to be made into a product like cheese.

Before the invention of refrigeration, cheese was the only dairy product that could be easily transported across long distances. Over the centuries, many distinctive types of hard cheeses have become associated with various regions of the world, such as Cheddar from England, Edam and Gouda from Holland, Gruyere from Switzerland, and Parmesan and Provolone from Italy.

A real dairy products industry has developed only in the last century or so, with the development of refrigeration and various kinds of specialized processing machinery, the scientific study of cattle breeding, and improved road and rail transportation systems for distributing manufactured products. The rise in urban populations also gave an extra impetus to the growth of the industry, as more and more people moved away from farm sources of dairy products and into cities.

Another important development was the introduction of pasteurization, named for the noted French chemist Louis Pasteur (1822-95). Many harmful bacteria can live in fresh milk. In the 1860s, Pasteur developed the process of pasteurization, which involves heating a foodstuff to a certain temperature for a specified period of time to kill the bacteria, then cooling the food again.

The Job

Dairy products manufacturing workers handle a wide variety of machines that process milk, manufacture dairy products, and prepare the products for shipping. Workers are usually classified by the type of machine they operate. Workers at some plants handle more than one type of machine.

Whole milk is delivered to a dairy processing plant from farms in large containers or in special tank trucks. The milk is stored in large vats until *dairy processing equipment operators* are ready to use it. First, the operator connects the vats to processing equipment with pipes, assembling whatever valves, bowls, plates, disks, impeller shafts, and other parts are needed to prepare the equipment for operation. Then the operator turns valves to pump a sterilizing solution and rinse water throughout the pipes and equipment. While keeping an eye on temperature and pressure gauges, the operator opens other valves to pump the whole milk into a centrifuge where it is spun at high speed to separate the cream from the skim milk. The milk is also pumped through a homogenizer to produce a specified emulsion (consistency that results from the distribution of fat through the milk) and, last, through a filter to remove any sediment. All this is done through continuous-flow machines.

The next step for the equipment operator is pasteurization, or the killing of bacteria that exist in the milk. The milk is heated by pumping steam or hot water through pipes in the pasteurization equipment. When it has been at the specified temperature for the correct length of time, a refrigerant is pumped through refrigerator coils in the equipment, which quickly brings the milk temperature down. Once the milk has been pasteurized, it is either bottled in glass, paper, or plastic containers, or it is pumped to other storage tanks for further processing. The dairy processing equipment operator may also add to the milk specified amounts of liquid or powdered ingredients, such as vitamins, lactic culture, stabilizer, or neutralizer, to make products such as buttermilk, yogurt, chocolate milk, or ice cream. The batch of milk is tested for acidity at various stages of this process, and each time the operator records the time, temperature, pressure, and volume readings for the milk. The operator may clean the equipment before processing the next batch of whole milk.

Processed milk includes a lot of nonfat dry milk, which is far easier to ship and store than fresh milk. Dry milk is produced in a gas-fired drier tended by a *drier operator*. The drier operator first turns on the equipment's drier mechanism, vacuum pump, and circulating fan and adjusts the flow controls. Once the proper drier temperature is reached, a pump sprays liquid milk into the heated vacuum chamber where milk droplets dry to powder and fall to the bottom of the chamber. The drier operator tests the dried powder for the proper moisture content and the chamber walls for burnt scale, which indicates excessive temperatures and appears as undesirable sediment when the milk is reconstituted. Milk-powder grinders operate equipment that mills and sifts the milk powder, ensuring a uniform texture.

For centuries, butter was made by hand in butter churns in which cream was agitated with a plunger until pieces of butter congealed and separated from the milk. Modern butter-making machines perform the same basic operation on a much larger scale. After sterilizing the machine, the *butter maker* starts a pump that admits a measured amount of pasteurized cream into the churn. The butter maker activates the churn and, as the cream is agitated by paddles, monitors the gradual separation of the butter from the milk. Once the process is complete, the milk is pumped out and stored, and the butter is sprayed with chlorinated water to remove excess remaining milk. With testing apparatus, the butter maker determines the butter's moisture and salt content and adjusts the consistency by adding or removing water. Finally, the butter maker examines the color and smells and tastes the butter to grade it according to predetermined standards.

In addition to the churn method, butter can also be produced by a chilling method. In this process, the butter maker pasteurizes and separates cream to obtain butter oil. The butter oil is tested in a standardizing vat for its levels of butter fat, moisture, salt content, and acidity. The butter maker adds appropriate amounts of water, alkali, and coloring to the butter oil and starts an agitator to mix the ingredients. The resulting mix is chilled in a vat at a specified temperature until it congeals into butter.

Cheese makers cook milk and other ingredients according to formulas to make cheese. The cheese maker first fills a cooking vat with milk of a pre-scribed butterfat content, heats the milk to a specified temperature, and dumps in measured amounts of dye and starter culture. The mixture is agi-tated and tested for acidity, which affects the rate at which enzymes coagu-late milk proteins and produce cheese. When a certain level of acidity has been reached, the cheese maker adds a measured amount of rennet, a sub-stance containing milk-curdling enzymes. The milk is left alone to coagulate into curd, the thick, protein-rich part of milk used to make cheese. The cheese maker later releases the whey, the watery portion of the milk, by pulling curd knives through the curd or using a hand scoop. Then the curd is agitated in the vat and cooked for a period of time, with the cheese maker squeezing and stretching samples of curd by hand and adjusting the cooking time to achieve the desired firmness or texture. Once this is done, the cheese maker or a cheese maker helper drains the whey from the curd, adds ingre-dients such as seasonings, and then molds, packs, cuts, piles, mills, and presses the curd into specified shapes. To make certain types of cheese, the curd may be immersed in brine, rolled in dry salt, pierced or smeared with a culture solution to develop mold growth, or placed on shelves to be cured. Later, the cheese maker samples the cheese for its taste, smell, look, and feel. Sampling and grading is also done by cheese graders, experts in cheeses who are required to have a state or federal license.

The distinctive qualities of various kinds of cheeses depend on a num-ber of factors, including the kind and condition of the milk, the cheesemak-ing process, and the method and duration of curing. For example, cottage cheese is made by the method described above. However, the cottage-cheese maker starts the temperature low and slowly increases it. When the curd reaches the proper consistency, the cottage-cheese maker stops the cooking process and drains off the whey. This method accounts for cottage cheese's loose consistency. Cottage cheese and other soft cheeses are not cured like hard cheeses and are meant for immediate consumption.

Process cheese products are made by blending and cooking different cheeses, cheese curd, or other ingredients such as cream, vegetable shorten-ing, sodium citrate, and disodium phosphate. The process-cheese cooker dumps the various ingredients into a vat and cooks them at a prescribed tem-perature. When the mixture reaches a certain consistency, the cooker pulls a lever to drain the cheese into a hopper or bucket. The process cheese may be pumped through a machine that makes its texture finer. Unheated cheese or curd may be mixed with other ingredients to make cold pack cheese or cream cheese. Other cheese workers include *casting-machine operators,* who tend the machines that form, cool, and cut the process cheese into slices of a specified size and weight, and grated-cheese makers, who handle the grinding, drying, and cooling equipment that makes grated cheese.

Ice cream is usually made from milk fat, nonfat milk solids, sweeteners, stabilizer (usually gelatin), and flavorings such as syrup, nuts, and fruit. Ice cream can be made in individual batches by batch freezers or in continuous-mix equipment by *freezer operators*. In the second method, the freezer operator measures the dry and liquid ingredients, such as the milk, coloring, flavoring, or fruit puree, and dumps them into the flavor vat. The mix is blended, pumped into freezer barrels, and injected with air. In the freezer barrel, the mix is agitated and scraped from the freezer walls while it slowly hardens. The operator then releases the ice cream through a valve outlet that may inject flavored syrup for rippled ice cream. The ice cream is transferred to a filling machine that pumps it into cartons, cones, cups, or molds for pies, rolls, and tarts. Other workers may process the ice cream into its various types, such as cones, varicolored packs, and special shapes. These workers include decorators, novelty makers, flavor room workers, and sandwich-machine operators.

Newly hired inexperienced workers in a dairy processing plant may start out as dairy helpers, cheese maker helpers, or cheese-making laborers. Beginning workers may do any of a wide variety of support tasks, such as scrubbing and sterilizing bottles and equipment, attaching pipes and fittings to machines, packing cartons, weighing containers, and moving stock. If they prove to be reliable, workers may be given more responsibility and assigned tasks such as filling tanks with milk or ingredients, examining canned milk for dirt or odor, monitoring machinery, cutting and wrapping butter and cheese, or filling cartons or bags with powdered milk. In time, workers may be trained to operate and repair any of the specialized processing machines found in the factory.

The raw milk at a dairy processing plant is supplied by *dairy farmers,* who raise and tend milk-producing livestock, usually cows. Dairy farmers often own their own farms, breed their own cows, and use special equipment to milk the cows, often twice a day. Many also perform other farm-related tasks, including growing crops. Assisting the dairy farmer is often the *dairy herd supervisor,* who takes milk samples from cows and tests the milk samples for information such as the amount of fat, protein, and other solids present in the milk. The dairy herd supervisor helps the farmer make certain that each cow in the herd is healthy and that the milk they produce will be fit for human consumption. Dairy herd supervisors do not generally work for one dairy farmer, but rather may oversee the milk production at a number of farms.

Requirements

High School

Most dairy products manufacturing workers learn their skills from company training sessions and on-the-job experience. Employers prefer to hire workers with at least a high school education. Courses that can provide helpful background for work in this field include mathematics, biology, and chemistry. Machine shop classes also can be useful for the experience gained in handling and repairing heavy machinery.

Postsecondary Training

Students interested in becoming cheese makers may find it necessary to obtain a college degree in a food technology or food science program. Dairy herd supervisors, in addition to a two-year or four-year food technology or food science degree, should try to gain experience working on a dairy farm. Those who seek management positions may need a bachelor's degree in food manufacturing with an emphasis on accounting, management, and other business courses.

Certification or Licensing

To ensure that consumers are receiving safe, healthful dairy foods, many dairy products manufacturing workers must be licensed by a state board of health or other local government unit. Licensing is intended to guarantee workers' knowledge of health laws, their skills in handling equipment, and their ability to grade the quality of various goods according to established standards. Some workers, such as cheese graders, may need to be licensed by the federal government as well.

Other Requirements

Many dairy manufacturing workers must pass physical examinations and receive certificates stating they are free from contagious diseases. An interest in food products and manual dexterity in operating equipment are important characteristics for this work.

Exploring

People who think they may be interested in working in the dairy products manufacturing industry may be able to find summer jobs as helpers in dairy processing plants. Assisting or at least observing equipment operators, cheese makers, butter makers, and others as they work is a good way to learn about this field. High school students may also find part-time or summer employment at dairy farms.

Starting Out

A good place to find information about job openings is at the personnel offices of local dairy processing plants. Other sources of information include newspaper classified ads and the offices of the state employment service. Those with associate's or bachelor's degrees in food technology, food science, or a related program can apply directly to dairy processing plants; many schools offering such programs provide job placement assistance. Dairy farmers often begin their careers by working on their own family farms.

Advancement

After gaining some experience, dairy products manufacturing workers may advance to become shift supervisors or production supervisors. Advancement is usually limited to those with at least an associate's or bachelor's degree in a food technology, food science, or related course of study. Formal training in related fields is necessary in order to move up to such positions as laboratory technician, plant engineer, or plant manager.

Workers who wish to change industries may find that many of their skills can be transferred to other types of food processing. With further training and education, they may eventually become dairy plant inspectors or technicians employed by local or state health departments.

Earnings

Earnings of dairy products manufacturing workers vary widely according to the responsibilities of the worker, geographical location, and other factors. Entry-level and unskilled workers can expect to begin at salaries around $15,000 per year. Dairy production workers with experience averaged approximately $27,500 per year in 1995. The overall average earnings for dairy production workers also varies according to the type of product produced by the plant. Workers processing fluid milk earned an average of $28,700 per year, while those at cheese processing plants averaged about $24,400 per year. Cheese makers and others with food technology degrees may earn as much as $40,000 per year or more. Production supervisors, plant engineers, and plant managers can earn $30,000 per year or more. Annual income for dairy farmers can vary widely, from as low as $10,000 per year to $90,000 per year and more; most dairy farmers own their own businesses and are responsible for its upkeep, as well as paying employees' salaries.

Dairy herd supervisors are paid based on the number of herd they test. Starting supervisors earn between $18,700 and $25,000, while experienced supervisors may earn $30,000 per year. The most experienced and highly trained supervisors can earn as much as $35,000 per year. Most dairy products workers are eligible for overtime pay for hours worked over 40 hours in a week. Benefits vary according to the company and its location, but sometimes include health insurance and vacation and sick pay.

Work Environment

Because of the strict health codes and sanitary standards to which they must adhere, dairy plants are generally clean, well-ventilated workplaces, equipped with modern, well-maintained machines. When workplace safety rules are followed, dairy processing plants are not hazardous places to work.

Workers in this industry generally report for work as early as 6 AM, with shifts ending around three in the afternoon. Dairy farmers and others may start work as early as four or five in the morning. People involved in the agriculture industry often work very long hours, often more than 12 hours per day. Many dairy products manufacturing workers stand during most of their workday. In some positions the work is very repetitive. Although the milk itself is generally transported from tank to tank via pipelines, some workers have to lift and carry other heavy items, such as cartons of flavoring, emulsifier, chemical additives, and finished products like cheese. To clean vats and other equipment, some workers have to get inside storage tanks and spray the walls with hot water, chemicals, or live steam.

Outlook

In 1996 there were 13,000 workers employed in this industry. The demand for American dairy products will probably remain high in the foreseeable future. Among the products that have grown in popularity in recent years are cheeses, ice cream, and lowfat milk. Despite this demand, employment in the dairy products manufacturing industry is expected to decline between 1996 and 2006. Improvements in technology and increased automation are two important factors contributing to this trend. However, because the milk industry is rarely affected by recessions or other economic difficulties and trends facing other industries, employment remains relatively stable and this industry suffers fewer layoffs than others. Because of continuing advances in the technology of dairy manufacturing and food science, the demand for laboratory technicians, plant engineers, and other technical staff is expected to remain strong.

For More Information

For information about the dairy industry, contact:

American Dairy Association
O'Hare International Centre
10255 West Higgins Road, Suite 900
Rosemont, IL 60018-2000
Tel: 847-803-2000

The following organization has a journal, a student affiliate, and information about the dairy industry.

American Dairy Science Association
1111 N. Dunlap Avenue
Savoy, IL 61874
Tel: 217-356-3182
Email: adsa@assochq.org
Web: http://www.adsa.uiuc.edu

The following is a federation of the Milk Industry Foundation, National Cheese Institute, and the International Ice Cream Association.

International Dairy Foods Association
1250 H Street, NW, Suite 900
Washington, DC 20005
Tel: 202-737-4332
Web: http://www.idfa.org

The following organization is concerned with milk quality and standards, animal health and food safety issues, dairy product labeling and standards, and legislation affecting the dairy industry.

National Milk Producers Federation
2101 Wilson Boulevard, Suite 400
Arlington, VA 22201
703-243-6111
Email: nmpf@aol.com
Web: http://www.nmpf.org

General Maintenance Mechanics

Overview

General maintenance mechanics, sometimes called maintenance technicians, repair and maintain machines, mechanical equipment, and buildings, and work on plumbing, electrical, and controls. They also do minor construction or carpentry work and routine preventive maintenance to keep the physical structures of businesses, schools, factories, and apartment buildings in good condition. They also maintain and repair specialized equipment and machinery found in cafeterias, laundries, hospitals, offices, and factories.

History

Before machines came to dominate the manufacturing of goods, craftsworkers had to learn many different kinds of skills. Blacksmiths, for example, had to know about forging techniques, horseshoeing, making decorative metalwork, and many other aspects of their trade. Carriage makers had to be familiar with carpentry, metalworking, wheel-making, upholstering, and design.

The Industrial Revolution set in motion many new trends, however, including a shift toward factory-type settings with workers who specialized in specific functions. This shift occurred partly because new machine production methods required a high degree of discipline and organization. Another reason for the change was because the new technology was so complex, no one person could be expected to master a whole field and keep up with changes that developed in it.

In a way, today's general maintenance mechanics recall craft workers of the era before specialization. They are jacks-of-all-trades. Typically they have a reasonable amount of skill in a variety of fields, including construction, electrical work, carpentry, plumbing, machining, direct digital controls, and sometimes other trades as well. They are responsible for keeping buildings and machines in good working order. In order to do this, they must have a broad understanding of mechanical tools and processes as well as the ability to apply their knowledge to solving problems. Today there are over a million general maintenance mechanics employed in the United States, working in almost every industry.

The Job

General maintenance mechanics perform almost any task that may be required to maintain a building or the equipment in it. They may be called on to replace faulty electrical outlets; fix air-conditioning motors; install water lines; build partitions; patch plaster or drywall; open clogged drains; dismantle, clean, and oil machinery; paint windows, doors, and woodwork; repair institutional-size dishwashers or laundry machines; and see to many other problems. Because of the diverse nature of the responsibilities of maintenance mechanics, they have to know how to use a variety of materials and be skilled in the use of most hand tools and ordinary power tools. They also must be able to recognize when they cannot handle a problem and must recommend that a specialized technician be called.

General maintenance mechanics work in many kinds of settings. Mechanics who work primarily on keeping industrial machines in good condition may be called *factory maintenance workers* or *mill maintenance workers,* while those mechanics who concentrate on the maintenance of a building's physical structure may be called *building maintenance workers* or *technicians.*

Once a problem or defect has been identified and diagnosed, maintenance mechanics must plan the repairs. They may consult blueprints, repair manuals, and parts catalogs to determine what to do. They obtain supplies and new parts from a storeroom or order them from a distributor. They install new parts in place of worn or broken ones, using hand tools, power tools, and sometimes electronic test devices and other specialized equipment. In some situations, maintenance mechanics may fix an old part or even fabricate a new part. To do this, they may need to set up and operate machine tools, such as lathes or milling machines, and operate gas- or arc-welding equipment to join metal parts together.

One of the most important kinds of duties general maintenance mechanics perform is routine preventive maintenance to correct defects before machinery breaks down or a building begins to deteriorate. This type of maintenance keeps small problems from turning into large, expensive ones. Mechanics often inspect machinery on a regular schedule, perhaps following a checklist that includes such items as inspecting belts, checking fluid levels, replacing filters, oiling moving parts, and so forth. They keep records of the repair work done and the inspection dates. Repair and inspection records can be important evidence of compliance with insurance requirements and government safety regulations.

New buildings often have computer-controlled systems, so mechanics who work in them must have basic computer skills. For example, newer buildings might have light sensors that are electronically controlled and automatically turn lights on and off. The maintenance mechanic has to understand how to make adjustments and repairs.

In small establishments, one mechanic may be the only person working in maintenance, and thus may be responsible for almost any kind of repair. In large establishments, however, tasks may be divided among several mechanics. For example, one mechanic may be assigned to install and set up new equipment, while another may handle preventive maintenance.

Requirements

High School

Many employers prefer to hire helpers or mechanics who are high school graduates, but a diploma is not always required. High school courses that provide good preparation for this occupation include mechanical drawing, metal shop, electrical shop, woodworking, blueprint reading, general science, and applied mathematics.

Postsecondary Training

Some mechanics learn their skills by working as helpers to people employed in building trades, such as electricians or carpenters. Other mechanics attend trade or vocational schools that teach many of the necessary skills. Becoming fully qualified for a mechanic's job usually requires one to four years of on-the-job training or classroom instruction, or some combination of both.

There are some certification and training programs open to maintenance mechanics. The BOMI Institute, for example, offers the designation of Systems Maintenance Technician (SMT) to applicants who have completed courses in boilers, heating systems, and applied mathematics; refrigeration systems and accessories; air handling, water treatment and plumbing systems; electrical and illumination systems; and building control systems. Technicians who have achieved SMT status can go on and become certified as Systems Maintenance Administrators (SMAs) by taking further classes in building design and maintenance, energy management, and supervision. While not necessarily required for employment, employees with certification may become more valuable assets to their employers and may have better chances at advancement.

Other Requirements

General maintenance mechanics need to have good manual dexterity and mechanical aptitude. People who enjoy taking things apart and putting them back together are good candidates for this position. Since some of the work, such as reaching, squatting, and lifting, requires physical strength and stamina, reasonably good health is necessary. Mechanics also need the ability to analyze and solve problems and to work effectively on their own, without constant supervision.

Exploring

Shop classes can give you a good indication of your mechanical aptitude and of whether or not you would enjoy maintenance work. The best way to experience the work these mechanics do, however, is to get a summer or part-time job as a maintenance helper in a factory, apartment complex, or similar setting. If such a job is not available, you might try talking with a maintenance mechanic to get a fuller, more complete picture of his or her responsibilities.

Employers

General maintenance mechanics are employed in factories, hospitals, schools, colleges, hotels, offices, stores, malls, gas and electric companies, government agencies, and apartment buildings throughout the United States. Statistics from the U.S. Bureau of Labor Statistics indicate that there are over 1 million people in the field. More than one-third are employed in service industries, such as elementary and secondary schools, colleges and universities, hospitals and nursing homes, and hotels. Others are employed in manufacturing industries, office and apartment buildings, government agencies, and utility companies.

Starting Out

General maintenance mechanics usually start as helpers to experienced mechanics and learn their skills on the job. Beginning helpers are given the simplest jobs, such as changing light bulbs or making minor drywall repairs. As general maintenance mechanics acquire skills, they are assigned more-complicated work, such as troubleshooting malfunctioning machinery.

Job seekers in this field usually apply directly to potential employers. Information on job openings for mechanic's helpers can often be found through newspaper classified ads, school placement offices, and the local offices of the state employment service. Graduates of trade or vocational schools may be able to get referrals and information from their school's placement office. Union offices may also be a good place to learn about job opportunities.

Advancement

Some general maintenance mechanics who are employed in large organizations may advance to supervisory positions. Another possibility is to move into one of the traditional building trades and become a craft worker, such as a plumber or electrician. In smaller organizations, promotion opportunities are limited, although increases in pay may result from an employee's good performance and increased value to the employer.

Earnings

Earnings for general maintenance mechanics vary widely depending on skill, geographical location, and industry. According to the BOMI Institute, the average annual wage for general maintenance mechanics in smaller facilities was about $31,000 in 1997. However, figures from the U.S. Bureau of Labor Statistics put the average wage at $24,400. The difference is probably because the BOMI Institute reaches maintenance mechanics who have more training. Those who work in large facilities with maintenance budgets of over $3 million earn on average $46,300 annually.

Almost all maintenance mechanics receive a benefits package that includes health insurance, paid vacation, sick leave, and a retirement plan. Mechanics earn overtime pay for work in excess of 40 hours per week.

Work Environment

General maintenance mechanics work in almost every industry and in a wide variety of facilities. In most cases, they work a 40-hour week. Some work evening or night shifts or on weekends; they may also be on call for emergency repairs. In the course of a single day, mechanics may do a variety of tasks in different parts of a building or in several buildings, and they may encounter different conditions in each spot. Sometimes they have to work in hot or cold conditions, on ladders, in awkward or cramped positions, among noisy machines, or in other uncomfortable places. Sometimes they must lift heavy weights. On the job, they must stay aware of potential hazards such as electrical shocks, burns, falls, and cuts and bruises. By following safety regulations and using tools properly, they can keep such risks to a minimum.

The mechanic who works in a small establishment may be the only maintenance worker and is often responsible for doing his or her job with little direct supervision. Those who work in larger establishments usually report to a maintenance supervisor who assigns tasks and directs their activities.

Outlook

The job outlook for general maintenance mechanics is good. Over the next several years, employment of these workers is expected to grow about as fast as average through 2006. Although the rate of construction of new apartment and office buildings, factories, hotels, schools, and stores is expected to be slower than in the past, most of these facilities still require the services of maintenance mechanics. In addition to newly created jobs, many openings will arise as experienced mechanics transfer to other occupations or leave the labor force.

General maintenance mechanics who work for manufacturing companies may be subject to layoffs during bad economic times, when their employers are under pressure to cut costs. Most mechanics, however, are not usually as vulnerable to layoffs related to economic conditions.

For More Information

For information on becoming certified as a systems maintenance technician or systems maintenance administrator, contact:

BOMI Institute
1521 Ritchie Highway
Arnold, MD 21012
Tel: 410-974-1417 or 800-235-BOMI

For information on general maintenance careers in building maintenance and construction, contact:

Mechanical Contractors Association of America
1385 Piccard Drive
Rockville, MD 20850
Tel: 301-869-5800
Web: http://www.mcaa.org/careers

Glass Manufacturing Workers

School Subjects
Chemistry
Technical/shop

Personal Skills
Technical/scientific
Mechanical/manipulative

Minimum Education Level
High school diploma

Salary Range
$28,300 to $32,000 to $33,600

Certification or Licensing
None available

Outlook
About as fast as the average

Overview

About 127,000 people work in the glass manufacturing industry, being employed as production workers in factories and plants that make glass and glass products from raw materials. These products include flat glass, such as window and plate glass; pressed glass items, such as glass dishes; blown glass items, such as light bulbs and many kinds of bottles; and various special products, such as glass blocks used in building construction, safety glass windshields, and glass for optical instruments.

History

People have been manufacturing glass for about 4,500 years. The earliest glass objects were produced taking little advantage of the special qualities of hot glass. A major advance came around 200 BC, when techniques were devised (probably in Syria) for blowing air into gobs of molten glass to shape the glass into useful objects. The new knowledge about working with hot glass spread quickly among glassmakers, and soon other peoples, notably the Romans, were making blown glass items. With the decline of Rome, much of the knowledge of working with glass was lost, not to be revived until glassworkers in Venice created a thriving industry around the 13th century. For hundreds of years, Venice was the leading center of glass production. In time, Venetian methods spread, new kinds of glass were developed, and good-quality blown glass was produced across much of Europe.

Skilled glassmakers were among the early European colonists in North America. However, not until the 18th century did glassmaking become a successful industry in the United States. Even at that time, glass was still made by hand and was so difficult to produce that it was expensive and seldom found in poor homes. In the 19th century, a steady stream of technological innovations simplified the various methods of production and made glass much more common. More efficient furnaces melted raw materials much faster; new molds made bottles much easier to mass produce; improved methods simplified the production of flat glass; and better polishing equipment greatly increased the output of plate glass for windows, creating a new look in buildings.

Major advances in the scientific understanding of glass and its properties have brought changes in manufacturing processes and new applications for glass products. Many new types of glass—such as heat-resistant glass, glass fabrics, and laminated glass—have been introduced. Although some craftworkers and artists still follow the old ways of making glass by hand, most modern glass is made in factories that use highly organized, automated industrial processes.

The Job

Glass manufacturing involves a number of basic operations, including mixing and melting materials; forming molten glass by blowing, pressing, casting, drawing, or rolling; heat-conditioning and controlled cooling; and finishing glass by polishing, coating, and using other surface processes. Different kinds of glass may involve different processes and require specialized workers. Most of today's glass manufacturing workers tend specialized machines used as part of a continuous mechanized operation.

Glass is usually made from sand (silicon dioxide), limestone (calcium carbonate), soda ash (sodium carbonate), and other raw ingredients. In many plants where glass is made, mixers tend equipment that blends ingredients. They either weigh and mix batches of materials or monitor machines that automatically supply the correct mix for melting in furnaces. *Cullet crushers* tend machines that crush and wash cullet, or broken waste glass, which will be recycled and melted with the raw ingredients. In some plants, batch-and-furnace operators control automatic equipment that can weigh and mix ingredients, then dump them into a furnace. *Combustion analysts* test and regulate the temperature of the furnace to manufacturing specifications. When the temperature is properly controlled, bubbles and impurities can be eliminated.

Many workers are concerned with machine-forming of the hot glass so they can take advantage of glass's malleable quality. Among these workers are *forming machine operators,* who set up and operate machines that press, blow, or spin lumps of molten glass into molds to make a wide variety of glass products, such as bottles, containers, and cathode-ray tubes. Under operator control, the machines deliver gobs of hot glass from the supply emerging from the furnace. Often, a puff of air is used to blow the glass firmly into a mold. The glass temperature is regulated until the molded item is ejected for further processing. *Pressers* tend press molds that force molten glass into shapes, making cast glassware items such as plates and automobile headlights. Others tend machines that extrude fiberglass filaments, mold optical glass blanks, form bulbs, and shape other glass products.

Flat glass is an extremely important product for windows, doors, and many other items. The float-glass process is used to produce much of the flat glass made today. In this process, molten glass flows from the furnace where it has been heated onto the surface of a pool of molten tin. The result is a glass with a good polish and flatness that requires less costly finish processing than other flat glass.

The glass manufacturing workers who make flat glass by other methods include *drawing-kiln operators,* who operate machines that process molten glass into continuous sheets by drawing molten glass upward from a tank and cooling it before it runs and loses its shape. Sometimes sheets of glass are made by *rolling-machine operators,* who operate equipment that rolls molten glass flat.

Some workers form hot glass by hand. They include *glassblowers,* who shape gobs of molten glass into glassware by blowing through a blowpipe, in much the same way that glass has been blown for centuries. Glassblowers produce certain kinds of special scientific equipment, as well as unique tableware and art objects. Other craftworkers shape and attach hot glass to other objects to make handles and pedestals.

Some glass is further processed with controlled reheating and slow cooling to eliminate flaws and internal stresses. *Lehr tenders* operate lehrs, which are tunnel-like automatic ovens used to heat-treat flat glass and glassware and fuse painted designs on glass.

Many glass products are not complete until they have been given other finishing treatments. Among the workers who do these tasks are *glass decorators,* who etch or cut designs into the surface of glass articles. *Glass grinders* remove rough edges and surface irregularities from glassware using belt or disk grinders. *Polishers* polish the edges and surfaces of flat glass, using polishing wheels.

Requirements

High School

Many workers in glass manufacturing occupations, such as machine tenders, can be hired as inexperienced beginners and learn the skills they need on the job. A good background for you if you plan to work in glass manufacturing would include high school courses in shop, general mathematics, and applied sciences. You might not be required to have a high school diploma, but many employers often prefer that you have graduated or at least have a GED (general equivalency diploma).

Postsecondary Training

Apprenticeship programs are recommended for training skilled glassmaking workers. These programs combine on-the-job training with formal instruction in related fields. Some apprenticeships are sponsored and run by local joint union-employer committees or by large glass manufacturing firms. The content of the training programs may vary somewhat, but programs usually last about three years. An example is the program of the Glass, Molders, Pottery, Plastics and Allied Workers International Union, which involves on-the-job work experience as well as classroom study.

Other Requirements

Because they work mostly with automated processes, glass manufacturing workers usually need only enough strength to lift light- or medium-weight objects. They must be able to tolerate repetitive work yet maintain careful attention to what they are doing while they oversee the operation of machines. Also, although union membership is not a requirement for employment (for example, most flat glass workers do not belong to a union), many workers in the glass manufacturing industry are represented by a union, such as the Glass, Molders, Pottery, Plastics and Allied Workers International Union and the American Flint Glass Workers Union.

Exploring

If you are interested in making glassware, art and shop courses in high school help you develop manual dexterity and learn about some of the tools and techniques that are used in glassmaking. Community art centers and adult education programs frequently offer classes in glassblowing, molding, and stained-glass construction. With the help of a teacher or guidance counselor, arrange to visit a glass manufacturing plant or a shop where artisans work with glass. One interesting field trip would take you to the Corning Museum of Glass at the Corning Glass Center in Corning, New York. The center is the third largest tourist site in the state of New York, and it offers educational information. The museum has 30,000 glass objects, from 3,500 years ago to the present; its library is the main research center for students of glass. And you can see a demonstration of actual glassblowing at the museum's Hot Glass Show.

Employers

Most workers in glass manufacturing work in factories in or near big cities in many sections of the country, and these work with pressed or blown glass. Others work in plants making glass containers, and some work with flat glass. One of the world leaders in specialty glass materials is Corning, the company that supplied the glass for Thomas Edison's first light bulb and influenced the use of red, yellow, and green lights for traffic control. Among the applications for Corning's glass technology were the first mass-produced TV tubes, freezer-to-oven ceramic cookware, and car headlights. In the 1970s, Corning pioneered the development of optical fiber and auto emission technology; in 1993 the company was chosen by AT&T to provide fiber-optic couplers for its undersea telecommunications system and developed an electrically heated catalytic converter that could beat strict California emissions standards. The company is the world's number one producer of tableware and cookware, led by its patented Pyrex and Corning Ware heat-resistant oven containers.

Starting Out

If you want an entry-level job in the glass manufacturing industry, you can apply directly to factories that may be hiring new workers. You might find leads to specific job openings through the classified ads in newspapers and the local offices of your state's employment service. Because many workers in this field are union members, it's a good idea to check out local union offices for job listings and general information about local opportunities.

If you want to be an apprentice in the industry, you might find information through union offices, glass manufacturing companies, and state services. After finishing your apprenticeship program, you could be rehired by the same company for which you apprenticed.

Advancement

Advancement opportunities for glassworkers are similar to those in many other fields. Maintaining discipline, motivation, and reliability in your job are often key to stepping up in seniority and earnings. You can be trained to operate many types of equipment, either through your company or with the help of your union. After you have gained some seniority and a diversity of glassmaking skills, you would be qualified to transfer to other jobs, shifts, or supervisory positions when they become available.

Earnings

Earnings of glass manufacturing workers depend on the type of industry they work in, their specific duties, union membership, the shift they work, and other factors. Production workers in flat-glass manufacturing average about $28,300 per year; those in pressed and blown glassmaking production jobs earn an average of $32,000 per year; those who produce glass containers average $33,600. Those who work over eight hours a day or 40 hours per week receive overtime pay, and they are usually paid at higher rates if they work at night, on weekends, or on holidays.

In many factories, the workers are union members, and their earnings are established according to agreements between the unions and company management. In addition, glassworkers often receive benefits, such as retirement plans and health and life insurance.

Work Environment

Glass factories usually operate around-the-clock, 24 hours every day of the year, because the furnaces have to be kept going all the time. For this reason, many workers work at night, on weekends, and on holidays. Although the standard workweek is about 40 hours, many workers put in overtime hours on a regular basis.

Factory conditions in glass plants have greatly improved over the years. On the job, workers may have to contend with some heat and fumes, but for the most part ventilation and heat shielding in modern plants have reduced worker exposure to these factors to very acceptable levels. Workers who tend furnaces and ovens are the most likely to work in hot conditions. Glass plants can be noisy, and workers may have to spend long periods of time on their feet.

Outlook

Glass is so common in our lives that as long as we continue to use it in its myriad forms, workers in glass manufacturing will be needed. It is difficult to say with any accuracy, however, whether job growth will be fast or slow or will remain the same. Much of the environment in the glass industry depends on other industries that use glass, like automobiles, spacecraft, nuclear energy, electronics, and solar energy. There are two markets for which new developments may be more important than others and thus require new workers: switchable glass (in which the ability of the glass to be seen through is changed by electronic and other means) and glass used in energy conservation.

For More Information

The following is a museum and educational site that has 30,000 objects of glass on view, a comprehensive library on the art and history of glass, and a studio that offers classes for all skill levels.

The Corning Museum of Glass
Education Department
One Museum Way
Corning, NY 14830-2253
Tel: 607-974-8257
Email: cmgeduc@servtech.com
Web: http://www.cmog.org/

This is a nonprofit trade association that represents the flat-glass industry and offers educational services, including the Auto Glass Certification Program, the Glass Management Institute, the Glazier Certification Program, the Auto Glass Technical Institute, and the Safety Management Program.

National Glass Association
8200 Greensboro Drive, Suite 302
McLean, VA 22102
Tel: 703-442-4890
Email: nga.glass@org
Web: http://www.glass.org

Industrial Chemicals Workers

Chemistry Mathematics	School Subjects
Following instructions Mechanical/manipulative	Personal Skills
Primarily indoors Primarily one location	Work Environment
High school diploma	Minimum Education Level
$21,000 to $38,000 to $50,000	Salary Range
None available	Certification or Licensing
Decline	Outlook

Overview

Industrial chemicals workers are employed in a variety of interrelated and interdependent industries and companies in which one concern often makes chemical precursors or starting materials for another's use. Most chemical workers convert the starting products or raw materials into other chemical compounds and derivative products, such as pharmaceuticals, plastics, solvents, and paints. In addition to being actively engaged in chemical operations, some workers are required to maintain safety, health, and environmental standards mandated by the federal government and perform routine and preventive maintenance tasks. Still others handle, store, and transport chemicals and operate batch processes.

History

Although its origins can be traced back to ancient Greece, chemistry was really recognized as a physical science during the 17th century. The alkali industry, which began then, made alkalis (caustic compounds such as sodium or potassium hydroxide) and alkaline salts such as soda ash (sodium carbonate) from wood and plant ashes. These compounds were then used to make soap and glass. By 1775, the natural sources of these alkaline compounds could not meet demand. Encouraged by the French Academy of Sciences, Nicholas Leblanc devised a synthetic process to manufacture them cheaply. Large-scale use of his process came a few years later in England. Inspired and encouraged by Leblanc's success, other scientists developed new methods for making a variety of industrially important chemicals. This marked the beginning of the modern industrial chemical industry. In the 1880s, the Leblanc process was superseded by the Solvay process. In the industrial chemical field today, many compounds such as ethylene, derived from petroleum, are used to synthesize countless other useful products. Ethylene winds up as polyethylene, polyethylene terephthalate, polystyrene, vinyl plastics, ethyl alcohol, and ethyl ether, to name just a few. Many of these, in turn, are used in fibers and fabrics, paints and resins, fuels, and pharmaceuticals. Thus you can see how one industry feeds off and relies on another. New uses for chemicals continue to be found, and new compounds to be synthesized, in research laboratories. Some of these compounds will eventually supplant those now in use. This is the natural order of things.

The Job

Workers in industrial chemical plants make all the products mentioned above plus thousands more. Basic chemicals such as sulfuric acid, nitric acid, hydrochloric acid, sodium hydroxide, sodium chloride, and ammonia are made by giant companies. The demand for these products is so great that only large companies can afford to build the factories and buy the equipment and the raw materials to produce these chemicals at the low prices for which they sell them. On the other hand, these giant companies rarely make specialty chemicals because they either can't afford to or they don't wish to make the necessary investment due to the very limited market. This is why small companies exist.

Because of the large variety of chemicals produced and the number of different processes involved, there are hundreds of job categories. Many of the jobs have quite a bit in common. In general, workers measure batches according to formulas; set reaction parameters for temperature, pressure, or flow of materials; and read gauges to monitor processes. They do routine testing, keep records, and may write progress reports. Many operators use computerized control panels to monitor processes. Some operate mixing machines, agitator tanks, blenders, steam cookers, and other pieces of equipment. Into a reaction vessel, the worker may pour two or more raw ingredients from other storage vats; empty cars from overhead conveyors, dumping the contents of a barrel or drum; or manually transfer materials from a hopper, box, or other container. The worker measures a preset amount of ingredients and then activates the mixing machine, while keeping an eye on the gauges and controls.

When the mixture has reached the desired consistency, color, or other characteristic, a test sample may be removed. If the analysis is satisfactory, the mixture is moved to its next destination either by piping, pumping into another container or processing machine, emptying into drums or vats, or by a conveyor. The operator then records the amount and condition of the mixture and readies his or her equipment for the next run.

Other workers may separate contaminants, undesirable byproducts, and unreacted materials with equipment that filters, strains, sifts, or centrifuges. Filters and centrifuges are often used to separate a slurry into liquid and solid parts. The *filter-press operator* sets up the press by covering the filter plates with canvas or paper sheets which separate the solids from the liquid portion. After the filtration, the plates are removed and cleaned. The centrifuge is a machine that spins a solid-liquid mixture like a washing machine in the spin cycle to separate it into solid and liquid components. If the desired end product is the liquid, the centrifuge operator discards the solids and vice versa.

Distillation operators use equipment that separates liquid mixtures by first heating them to their boiling points. The heated vapors rise into a distillation column. If a very pure liquid is desired, a fractional distillation column is used. A distillation apparatus consists of an electrically or steam-heated still pot, a distillation column, a water-cooled condenser, and a collector. In this process, the hot vapors rise through the distillation column. The condenser cools the vapors and converts them back into a liquid. The condensed liquid is collected and removed for further use. Distillation, a very important separation technique for purifying and separating liquids, is widely used in the liquor industry, petroleum refineries, and chemical companies that make and use liquid chemicals.

Solid chemical mixtures often need to be dried before they can be used. Workers heat, bake, dry, and melt chemicals with kilns, vacuum dryers, rotary or tunnel furnaces, and spray dryers. The workers who operate this equipment, regardless of the industry, perform the same operations.

The paint industry manufactures paints, varnishes, shellacs, lacquers, and a variety of liquid products for decorative and protective coatings. It not only makes many of the materials that go into its products but also purchases chemicals, resins, solvents, dyes, and pigments from others. In its operations, it performs many of the tasks described above. Coating and laminating are related industries. Their workers operate press rollers; laminating, coating, and printing machines; and sprayers. They carefully apply measured thicknesses of coating materials to a variety of substrates, such as paper, plastic, metal, and fabric.

Requirements

High School

Most of the equipment in the industrial chemical industry is now automated and computer controlled. Because of the complex equipment used, employers prefer to hire workers with a minimum of a high school diploma. Knowledge of basic mathematics, science, and computer skills is essential for those seeking employment in this field. Machine shop experience is also useful.

Postsecondary Training

Entry-level employees always get on-the-job training and special classroom work. Classes may include heat transfer principles, the basics of distillation, how to take readings on tanks and other equipment, and how to read blueprints. Workers also get safety training about the chemicals and processes they will encounter.

More advanced knowledge of chemistry and physics is important for those who hope to advance to supervisory and managerial positions. Training to become a skilled operator may take two to five years. Information on apprenticeship programs can be found through state employment bureaus. Some community colleges have study programs that allow students to combine classroom work with on-the-job experience to enhance their skills and knowledge.

Other Requirements

Workers in this industry must be dependable, alert, accurate, and able to follow instructions exactly. They must always be mindful of the potential hazards involved in working with chemicals and cannot ever be careless. They should be conscientious, be able to work without direct supervision, accept repetitive and sometimes monotonous work, and be willing to work with others.

Exploring

A helpful and inexpensive way to explore employment opportunities is to talk with someone who has worked in the industry in which you are interested. Also, it may be possible to arrange a tour of a manufacturing plant by contacting its public relations department. Another way to explore chemical manufacturing occupations is to check high school or public libraries for books on the industry. Other sources include trade journals, high school guidance counselors, and university placement offices. Students should join high school science clubs. Students can also subscribe to the American Chemical Society's *Chem Matters,* a quarterly magazine for high school chemistry students.

Employers

Industrial chemical workers are a necessary part of all chemical manufacturing whether the industry in question is producing basic chemicals, pharmaceuticals, paints, food, or a myriad of other products. The companies vary in size, depending on the nature of the products they produce. Some large industrial chemical companies—DuPont, Dow, and Union Carbide, for example—may make the chemicals they use in their own operations. Others purchase what they need from specialty chemical companies, such as Mallincrodt and J. T. Baker.

Basic chemicals, such as sodium hydroxide and nitric acid, are usually made by giant companies while small companies may make fine or specialty chemicals to supply to other manufacturers. Some of the duties are involved in the actual production process; others concern the equipment used in manufacturing; still others test finished products to insure that they meet industry and government standards of purity and safety. There are a number of government laboratories, such as the Department of Agriculture, the Bureau of Standards, and the Bureau of Mines, that employ chemical workers.

Starting Out

High school graduates qualify for entry-level factory jobs as helpers, laborers, or material movers. They learn how to handle chemicals safely and acquire skills that enable them to advance to higher levels of responsibility. Students interested in a job in the industrial chemical industry should look for information on job openings through classified ads and employment agencies. Information can also be obtained by contacting the personnel offices of individual chemical plants and local union offices of the International Chemical Workers Union and the Oil, Chemical, and Atomic Workers International Union. High school and college guidance and job placement offices are other knowledgeable sources.

Advancement

Movement into higher-paying jobs is possible with increasing experience and on-the-job training. Advancement usually requires mastery of advanced skills. Employers often offer classes for those who want to improve their skills and advance their careers.

Most workers start as laborers or unskilled helpers. They can advance to mechanic and installer jobs through formal vocational or in-house training. Or they can move up to positions as skilled operators of complex processes. They may become operators who monitor the flow and mix ratio of chemicals as they go through the production process. Experienced and well-trained production workers can advance to become supervisors overseeing an entire process.

Earnings

In 1997, median annual earnings for production workers in the industrial chemicals industry were $31,580 ($14.95 an hour) and for chemical plant and system operators, $37,380 ($18.27 an hour). Workers are usually paid more for night, weekend, and overtime work. Hourly rates for each production job are often set by union contract. Fringe benefits vary among employers. They may include group hospital, dental, and life insurance; paid holidays and vacations; and pension plans. Also, many workers qualify for college tuition aid from their companies.

Work Environment

Working conditions in plants vary, depending on specific jobs, the type and condition of the equipment used, and the size and age of the plant. Chemical processing jobs used to be very dangerous, dirty, and disagreeable. However, working conditions have steadily improved over the years as a result of environmental, safety, and health standards mandated by the government. As a result of government intervention, chemical manufacturing now has an excellent safety record that is superior to other manufacturing industries. Nevertheless, chemical plants by their very nature can be extremely hazardous if strict safety procedures are not followed and enforced. Precautions include wearing protective clothing and equipment where required. Hard hats and safety goggles are worn throughout the plant.

Although few jobs in this industry are strenuous, they may become monotonous. Since manufacturing is a continuing process, most chemical plants operate around the clock. Once a process has begun, it cannot be stopped. This means that workers are needed for three shifts—split, weekend, and night shifts are common.

Outlook

The industrial chemicals industry has seen little real growth in output and productivity since 1990, and output is expected to remain essentially flat through 2006. The industry has had to spend much money to comply with government pollution and safety regulations, thus increasing production costs and shrinking already small profit margins. Increasing competition with overseas chemical manufacturers is another factor that clouds the prospects of future growth for the domestic chemical industry.

The industrial chemical industry remains a good-sized employer with over 250,000 workers in the late 1990s. Going into the first years of the 21st century, the U.S. Department of Labor anticipates a small decline in job opportunities for plant operators and only a 4 percent increase in equipment workers, primarily because of more efficient production processes and increased plant automation. With the growing application of computerized controls comes less need for on-site personnel to oversee processes and operate equipment. On the other hand, advancing technology should create jobs for technical workers with the necessary skills to handle increasingly complex chemical processes and controls.

Since the industry provides good working conditions, wages, and fringe benefits, job turnover is low. Still, the Department of Labor projects there will be about 35,000 job openings through 2006, mostly to replace those who retire or otherwise leave the field.

For More Information

To subscribe to Chem Matters *or to learn more about chemical process industries and technical operators, contact:*

American Chemical Society (ACS)
Career Education
1155 16th Street, NW
Washington, DC 20036
Tel: 202-452-2113
Web: http://www.acs.org

Those interested in becoming process control engineers can receive career guidance materials from:

American Institute of Chemical Engineers (AIChE)
Communications Department
345 East 47th Street
New York, NY 10017-2395
Tel: 212-705-7660 or 800-242-4363
Web: http://www.aiche.org

Industrial Engineering Technicians

	School Subjects
Computer science Mathematics	
	Personal Skills
Communication/ideas Technical/scientific	
	Work Environment
Primarily indoors Primarily one location	
	Minimum Education Level
Associate's degree	
	Salary Range
$22,300 to $38,400 to $59,800	
	Certification or Licensing
Voluntary	
	Outlook
About as fast as the average	

Overview

Industrial engineering technicians assist industrial engineers in their duties: they collect and analyze data and make recommendations for the efficient use of personnel, materials, and machines to produce goods or to provide services. They may study the time, movements, and methods a worker uses to accomplish daily tasks in production, maintenance, or clerical areas.

Industrial engineering technicians prepare charts to illustrate work flow, floor layouts, materials handling, and machine utilization. They make statistical studies, analyze production costs, prepare layouts of machinery and equipment, help plan work flow and work assignments, and recommend revisions to revamp production methods or improve standards. As part of their job, industrial engineering technicians often use equipment such as computers, timers, and camcorders.

History

Industrial engineering is a direct outgrowth of the Industrial Revolution, which began in England in the 18th century and later spread to the United States. By linking a power source, such as a steam engine, to simple mechanical devices, early mechanical and industrial engineers were able to design and build factories to rapidly and economically produce textiles, clothing, and other materials.

Today, factories in the United States and around the world produce almost all of our consumer goods. This tremendous growth led to a need for industrial engineers, who evaluate not only the machines that go into the factory but also the raw materials, the people who run the machines, the costs and efficiency of operations, and other factors that affect the success of an industrial operation.

Industrial engineering as a separate specialty emerged during the 20th century. For as long as there have been industrial engineers, however, there have been skilled assistants who work with them and handle tasks that do not require the engineer's direct involvement. Today's industrial engineering technicians are the direct descendants of those assistants. As the years have gone by, the number, variety, and complexity of the responsibilities falling to industrial engineering technicians have increased greatly. In the past, assistants could rely purely on common sense and on-the-job experience, but today's industrial engineering technicians must be specially trained and educated before entering the workplace.

Today, the scope of industrial engineering goes far beyond the factory. The principles of work flow and quality control are now applied to other work environments, including corporations' offices and retail stores. The industrial engineering technician is recognized and respected as a team member in evaluating and improving production and working conditions.

The Job

The type of work done by an industrial engineering technician depends on the location, size, and products of the company for which he or she works. Usually a technician's duties fall into one or more of the following areas: work measurement, production control, wage and job evaluation, quality control, or plant layout.

Industrial engineering technicians involved in methods engineering analyze new and existing products to determine the best way to make them at the lowest cost. In these analyses, *methods engineering technicians* recommend which processing equipment to use; determine how fast materials can be processed; develop flowcharts; and consider all materials-handling, movement, and storage aspects of the production.

The *materials-handling technician* studies the current methods of handling material, then compares and evaluates alternatives. The technician will suggest changes that reduce physical effort, make handling safer, and lower costs and damage to products.

Work measurement technicians study the production rate of a given product and determine how much time is needed for all the activities involved. They do this by timing the motions necessary for a complete operation, analyzing tapes of workers, and consulting historical statistics collected in the factory. Time study technicians analyze and determine elements of work, their order, and the time required to produce a part.

The *production control technicians* often work in scheduling departments, where they coordinate many complex details to ensure product delivery on a specified date. To do this, production control technicians must know the products and assemblies to be made, the routes to be used through the plant, and the time required for the process. These technicians also issue orders to manufacture products, check machine loads, and maintain constant surveillance of the master schedules.

Production control technicians also work in dispatching offices, where they issue orders to the production areas, watch department machine loads, report progress of products, and expedite the delivery of needed parts to avoid delays.

Inventory control technicians maintain inventories of raw materials, semi-finished products, completed products, packaging materials, and supplies. They ensure an adequate supply of raw materials, watch for obsolete parts, and prevent damage or loss to products.

Quality control technicians work with inspection departments to maintain quality standards set by production engineers. They check all incoming materials and forecast the quality of obtainable materials. Quality control technicians use a variety of techniques and perform other duties that include part-drawing surveillance, checking of parts with inspection tools, identifying trouble, and providing corrective procedures.

Cost control technicians compare actual product costs with budgeted allowances. These technicians investigate cost discrepancies, offer corrective measures, and analyze results.

Budget technicians gather figures and facts to project and graph break-even points. They help prepare budgets for management and present the effects of production schedules on profitability.

Wage and job evaluation technicians gather and organize information pertaining to the skill, manual effort, education, and other factors involved in the jobs of all hourly employees. This information helps to set salary ranges and establish job descriptions.

The *plant layout technician* works with materials-handling personnel, supervisors, and management to help make alterations in manufacturing facilities. These technicians study old floor plans; consider all present and future aspects of operations; and revise, consult, and then propose layouts to production and management personnel.

Requirements

High School

Prospective industrial engineering technicians should take classes in algebra, geometry, calculus, chemistry, physics, trigonometry, and English. Mechanical drawing, metal shop, and communications would also be helpful. Computers have become the most used tool in industrial engineering, so computer science classes are critical to a student considering a career in this field. Also recommended are courses in shop sketching, blueprint reading, mechanical drawing, and model making, if available.

Postsecondary Training

Most employers prefer to hire someone with at least a two-year degree in engineering technology although it is possible to qualify for some jobs with no formal training. Training is available at technical institutes, junior and community colleges, extension divisions of universities, public and private vocational-technical schools, and through some technical training programs in the armed services.

Most two-year associate programs accredited by the Accreditation Board for Engineering and Technology (ABET) include in the first year courses in mathematics, orthographic and isometric sketching, blueprint reading, manufacturing processes, communications, technical reporting, introduction to numerical control, and introduction to computer-aided design (CAD).

Typical second-year courses include methods, operation, and safety engineering; industrial materials; statistics; quality control; computer control of industrial processes; plant layout and materials handling; process planning and manufacturing costs; production problems; psychology and human relations; and industrial organization and institutions. Since the type and quality of programs and schools vary considerably, prospective students are advised to consider ABET-accredited programs first.

Certification or Licensing

To give recognition and encouragement to industrial engineering technicians, the National Institute for Certification in Engineering Technologies (NICET) has established a certification program that some technicians may wish to consider. Although certification is not generally required by employers, those with certification often have a competitive advantage when it comes to hiring and promotions. Certification is available at various levels, each combining a written examination in a specialty field with a specified amount of job-related experience.

Other Requirements

Industrial engineering technicians should be adept at compiling and organizing data and be able to express themselves clearly and persuasively both orally and in writing. They should be detail oriented and enjoy solving problems.

Exploring

Opportunities to gain experience in high school are somewhat limited. However, students can obtain part-time work or summer jobs in industrial settings, even if not specifically in the industrial engineering area. Although this work may consist of menial tasks, it offers firsthand experience and demonstrates interest to future employers. Part-time jobs often lead to permanent employment, and some companies offer tuition reimbursement for educational costs.

Insights into the industrial engineering field can also be obtained in less direct ways. Professional associations regularly publish newsletters and other information relevant to the technician. Industrial firms frequently advertise or publish articles in professional journals or in business and general interest magazines that discuss innovations in plant layout, cost control, and productivity improvements. By finding and collecting these articles, prospective technicians can acquaint themselves with and stay informed on developments in the field.

Employers

Industrial engineering technicians are most often found in durable goods manufacturing, such as electronic and electrical machinery and equipment, industrial machinery and equipment, instruments, and transportation equipment. Some technicians are employed by engineering and business services companies that do contract engineering work. The U.S. Departments of Defense, Transportation, Agriculture, and Interior are also major employers, along with state and municipal governments.

Starting Out

Many industrial engineering technicians find their first jobs through interviews with company recruiters who visit campuses. In many cases, students are invited to visit the prospective employer's plant for further consultation and to become better acquainted with the area, product, and facilities. For many students, the job placement office of their college or technical school is the best source of possible jobs. Local manufacturers or companies are in constant contact with these facilities, so they have the most up-to-date job listings.

Advancement

As industrial engineering technicians gain additional experience, and especially if they pursue further education, they become candidates for advancement. Continuing education is fast becoming the most important way to advance. Many employers encourage this and will reimburse education costs.

The typical advancement path for industrial engineering technicians is to become a supervisor, an industrial engineer, or possibly a chief industrial engineer.

Here are some examples of positions to which technicians might aspire:

Production control managers supervise all production control employees, train new technicians, and coordinate manufacturing departments.

Production supervisors oversee manufacturing personnel and compare departmental records of production, scrap, and expenditures with departmental allowances.

Plant layout engineers supervise all plant-layout department personnel, estimate costs, and confer directly with other department heads to obtain information needed by the layout department.

Managers of quality control supervise all inspection and quality control employees, select techniques, teach employees new techniques, and meet with tool room and production people when manufacturing tolerances or scrap become a problem.

Chief industrial engineers supervise all industrial engineering employees, consult with department heads, direct departmental projects, set budgets, and prepare reports.

Earnings

The salary range for entry-level industrial engineering technicians varies according to the product being manufactured, geographic location, and the education and skills of the technician. According to the U.S. Bureau of Labor Statistics, the average annual salary for industrial engineer technicians in 1997 was $38,360. Some technicians, however, especially those at the very beginning of their careers, earn about $22,000 a year, while some senior technicians with special skills and experience earn over $59,000 a year. In addition to salary, most employers offer paid vacation time, holidays, insurance and retirement plans, and tuition assistance for work-related courses.

Work Environment

Industrial engineering technicians generally work indoors. Depending on their jobs, they may work in the shop or office areas or in both. The type of plant facilities depends on the product. For example, an electronics plant producing small electronic products requiring very exacting tolerances has very clean working conditions.

Industrial engineering technicians often travel to other locations or areas. They may accompany engineers to technical conventions or on visits to other companies to gain insight into new or different methods of operation and production.

Continuing education plays a large role in the life of industrial engineering technicians. They may attend classes or seminars to keep up-to-date with emerging technology and methods of managing production efficiently.

Hours of work may vary and depend on factory shifts. Industrial engineering technicians are often asked to get jobs done quickly and to meet very tight deadlines.

Outlook

As products become more technically demanding to produce, competitive pressures will force companies to improve and update manufacturing facilities and product designs. Thus, the demand for well-trained industrial engineering technicians will stay about average through 2006. Opportunities will be best for individuals who have up-to-date skills. As technology becomes more sophisticated, employers will continue to seek technicians who require the least amount of additional job training.

The employment outlook varies with area of specialization and industry. For example, changing and increasing numbers of environmental and safety regulations may lead companies to revise some of their procedures and practices, and new technicians may be needed to assist in these changeovers. However, technicians whose jobs are defense-related may experience fewer opportunities because of recent defense cutbacks.

Prospective technicians should keep in mind that advances in technology and management techniques make industrial engineering a constantly changing field. Technicians will be able to take advantage of new opportunities only if they are willing to continue their training and education throughout their careers.

For More Information

For information about membership in a professional society specifically created for engineering technicians, contact:

American Society of Certified Engineering Technicians (ASCET)
PO Box 1348
Flowery Branch, GA 30542
Tel: 770-967-9173

For more information on careers and training as an industrial engineering technician, contact:

IEEE Industry Applications Society
c/o Institute of Electrical and Electronics Engineers
3 Park Avenue, 17th Floor
New York, NY 10017
Tel: 212-419-7900
Web: http://www.ieee.org/eab/

Institute of Industrial Engineers
25 Technology Park/Atlanta
Norcross, GA 30092
Tel: 404-449-0460
Web: http://www.iienet.org

For more information on careers as an engineering technician, contact:

Junior Engineering Technical Society, Inc.
1420 King Street, Suite 405
Alexandria, VA 22314-2794
Tel: 703-548-5387
Web: http://www.asee.org

For information about obtaining certification, contact:

National Institute for Certification in Engineering Technologies
1420 King Street
Alexandria, VA 22314-2715
Tel: 888-476-4238
Web: http://www.nicet.org

Industrial Engineers

Computer science Mathematics	School Subjects
Leadership/management Technical/scientific	Personal Skills
Primarily indoors Primarily one location	Work Environment
Bachelor's degree	Minimum Education Level
$38,000 to $52,300 to $90,000	Salary Range
Required by certain states	Certification or Licensing
About as fast as the average	Outlook

Overview

Industrial engineers use their knowledge of various disciplines—including systems engineering, management science, operations research, and fields such as ergonomics—to determine the most efficient and cost-effective methods for industrial production. Engineers are responsible for designing systems that integrate materials, equipment, information, and people in the overall production process.

History

In today's industries, manufacturers increasingly depend on industrial engineers to determine the most efficient production techniques and processes. The roots of industrial engineering, however, can be traced to ancient Greece, where records indicate that manufacturing labor was divided among people having specialized skills.

The most significant milestones in industrial engineering—before the field even had an official name—occurred in the 18th century, when a number of inventions were introduced in the textile industry. The first was the flying shuttle that opened the door to the highly automatic weaving we now take for granted. This shuttle allowed one person, rather than two, to weave fabrics wider than ever before. Other innovative devices, such as the power loom and the spinning jenny that increased weaving speed and improved quality, soon followed. By the late 18th century, the Industrial Revolution was in full swing. Innovations in manufacturing were made, standardization of interchangeable parts was implemented, and specialization of labor was increasingly put into practice.

Industrial engineering as a science is said to have originated with the work of Frederick Taylor. In 1881, he began to study the way production workers used their time. At the Midvale Steel Company where he was employed, he introduced the concept of time study, whereby workers were timed with a stopwatch and their production was evaluated. He used the studies to design methods and equipment that allowed tasks to be done more efficiently.

In the early 1900s, the field was known as scientific management. Frank and Lillian Gilbreth were influential with their motion studies of workers performing various tasks. Then, around 1913, automaker Henry Ford implemented a conveyor belt assembly line in his factory, which led to increasingly integrated production lines in more and more companies. Industrial engineers nowadays are called upon to solve ever more complex operating problems and to design systems involving large numbers of workers, complicated equipment, and vast amounts of information. They meet this challenge by utilizing advanced computers and software to design complex mathematical models and other simulations.

The Job

Industrial designers are involved with the development and implementation of the systems and procedures that are utilized by many industries and businesses. In general, they figure out the most effective ways to use the three basic elements of any company: people, facilities, and equipment.

Although industrial engineers work in a variety of businesses, the main focus of the discipline is in manufacturing, also called industrial production. Primarily, industrial engineers are concerned with process technology, which includes the design and layout of machinery and the organization of workers who implement the required tasks.

Industrial engineers' responsibilities are numerous. With regard to facilities and equipment, engineers are involved in selecting machinery and other equipment and then in setting them up in the most efficient production layout. They also develop methods to accomplish production tasks, such as the organization of an assembly line. In addition, they devise systems for quality control, distribution, and inventory.

Industrial engineers are responsible for some organizational issues. For instance, they might study an organization chart and other information about a project and then determine the functions and responsibilities of workers. They devise and implement job evaluation procedures as well as articulate labor-utilization standards for workers. Engineers often meet with managers to discuss cost analysis, financial planning, job evaluation, and salary administration. Not only do they recommend methods for improving employee efficiency but they may also devise wage and incentive programs.

Industrial engineers evaluate ergonomic issues—the relationship between human capabilities and the physical environment in which they work. For example, they might evaluate whether machines are causing physical harm or discomfort to workers, or whether the machines could be designed differently to enable workers to be more productive.

In industries that do not focus on manufacturing, industrial engineers are often called management analysts or management engineers. In the health care industry, such engineers are asked to evaluate current administrative and other procedures. They also advise on job standards, cost-containment, and operations consolidation. Some industrial engineers are employed by financial services companies. Because many engineering concepts are relevant in the banking industry, engineers there design methods to optimize the ratio of tellers to customers and the use of computers for various tasks and to handle mass distribution of items such as credit card statements.

Requirements

High School

To prepare for a college engineering program, concentrate on mathematics (algebra, trigonometry, geometry, calculus), physical sciences (physics, chemistry), social sciences (economics, sociology), and English. Engineers often have to convey ideas graphically and may need to visualize processes in three-dimension, so courses in graphics, drafting, or design are also help-ful. Students should take honors level courses if possible.

Postsecondary Training

A bachelor's degree from an accredited institution is usually the minimum requirement for all professional positions. There are about 100 accredited industrial engineering programs in the United States. Colleges offer either four- or five-year engineering programs. Because of the intensity of the cur-ricula, many students take heavy course loads and attend summer sessions in order to finish in four years.

During their junior and senior years, students should be considering specific career goals, such as in which industry to work. Third- and fourth-year courses focus on such subjects as facility planning and design, work measurement standards, process design, engineering economics, manufac-turing and automation, and incentive plans.

Many industrial engineers go on for a postgraduate degree. These pro-grams tend to involve more research and independent study. Graduate degrees are usually required for teaching positions.

Certification or Licensing

Registration as a professional engineer is generally voluntary but is often con-sidered when employers are reviewing workers for promotion. Registration guidelines are different in each state. Normally they involve meeting certain educational requirements and passing an eight-hour Fundamentals of Engineering (FE) exam in your junior year. After working a specified amount of time in engineering, you must pass the eight-hour Professional Engineering (PE) exam. Another credential is the Systems Integration Certificate, which is offered by the Institute of Industrial Engineers to those who have been working in the field for at least five years.

Other Requirements

Industrial engineers enjoy problem solving and analyzing things as well as being a team member. The ability to communicate is vital since engineers interact with all levels of management and workers. Being organized and detail-minded is important because industrial engineers often handle large projects and must bring them in on time and on budget. Since process design is the cornerstone of the field, an engineer should be creative and inventive.

Exploring

Try joining a science or engineering club, such as the Junior Engineering Technical Society (JETS). JETS offers academic competitions in subjects such as computer fundamentals, mathematics, physics, and English. It also conducts design contests in which students learn and apply science and engineering principles. Membership in JETS includes a quarterly magazine, *JETS Report,* that has interviews and articles on various engineering careers. You also might read some engineering books for background on the field or magazines such as *Industrial Engineering.*

Other opportunities for exploring industrial engineering careers can be found at summer camps. For example, the Worcester Polytechnic Institute in Massachusetts offers a 12-day session for high school seniors that focuses on programs in science, math, and various engineering disciplines while offering recreational activities.

Employers

Although a majority of industrial engineers are employed in the manufacturing industry, related jobs are found in almost all businesses, including transportation; communications; electric; gas and sanitary services; government; finance; insurance; real estate; wholesale and retail trade; construction; mining; agriculture; forestry; and fishing. Also, many work as independent consultants.

Starting Out

The main qualification for an entry-level job is a bachelor's degree in industrial engineering. Accredited college programs generally have job openings listed in their placement offices. Entry-level industrial engineers find jobs in various departments, such as computer operations, warehousing, and quality control. As engineers gain on-the-job experience and familiarity with departments, they may decide on a specialty. Some may want to continue to work as process designers or methods engineers, while others may move on to administrative positions.

Some further examples of specialties include work measurement standards; shipping and receiving; cost control; engineering economics; materials handling; management information systems; mathematical models; and operations. Many who choose industrial engineering as a career find its appeal in the diversity of sectors that are available to explore.

Advancement

After having worked at least three years in the same job, an industrial engineer may have the basic credentials needed for advancement to a higher position. In general, positions in operations and administration are considered high-level jobs, although this varies from company to company. Engineers who work in these areas tend to earn larger salaries than those who work in warehousing or cost control, for example. If one is interested in moving to a different company, it is considered easier to do so within the same industry.

Industrial engineering jobs are often considered stepping stones to management positions—even in other fields. Engineers with many years' experience frequently are promoted to higher level jobs with greater responsibilities. Because of the field's broad exposure, industrial engineering employees are generally considered better prepared for executive roles than are other types of engineers.

Earnings

According to the U.S. Bureau of Labor Statistics, the average annual wage for industrial engineers in 1997 was $52,350. As with most occupations, salaries rise as more experience is gained. Veteran engineers can earn over $90,000. According to a survey by the National Association of Colleges and Employers, the average starting salary for industrial engineers is $38,000.

Salaries for industrial engineers also vary with regard to geographic location. Those who work in states along the Pacific, for example, tend to make more than those in any other area in the United States. Industrial engineers working in the Great Lakes area, such as in Illinois and Indiana, generally make less than those in the Pacific states but more than those in the north-central part of the country, where salaries in this field are usually lowest.

Work Environment

Industrial engineers usually work in offices at desks and computers, designing and evaluating plans, statistics, and other documents. Overall, industrial engineering is ranked above other engineering disciplines for factors such as employment outlook, salary, and physical environment. However, industrial engineering jobs are considered stressful because they often entail tight deadlines and demanding quotas, and jobs are moderately competitive. Engineers work an average of 46 hours per week.

Industrial engineers generally collaborate with other employees, conferring on designs and procedures, as well as with business managers and consultants. Although they spend most of their time in their offices, they frequently must evaluate conditions at factories and plants, where noise levels are often high.

Outlook

In 1997, industrial engineers held about 112,400 jobs in the United States. The U.S. Bureau of Labor Statistics anticipates that industrial engineering opportunities will grow through the year 2006; however, this increase will not be any faster than the average growth for all occupations.

Engineers who transfer or retire will create the highest percentage of openings in this field. New jobs will be found in growing industries, especially those implementing automation to solve complicated business operations. The demand for industrial engineers will continue as manufacturing and other companies strive to make their production processes more effective and competitive.

Although approximately 75 percent of industrial engineering jobs are in the manufacturing industry, opportunities are found in a diversity of fields because the skills involved can be used in nearly any type of business.

For More Information

For a list of ABET-accredited engineering schools, contact:

Accreditation Board for Engineering and Technology, Inc.
111 Market Place, Suite 1050
Baltimore, MD 21202-4012
Tel: 410-347-7700
Web: http://www.abet.ba.md.us

For information about colleges and careers in industrial engineering, contact:

Institute of Industrial Engineers
25 Technology Park
Norcross, GA 30092
Tel: 770-449-0461
Web: http://www.iienet.org

Industrial Machinery Mechanics

	School Subjects
Mathematics Technical/Shop	
	Personal Skills
Mechanical/manipulative Technical/scientific	
	Work Environment
Primarily indoors Primarily one location	
	Minimum Education Level
High school diploma	
	Salary Range
$17,000 to $31,600 to $43,000	
	Certification or Licensing
None available	
	Outlook
Little change or more slowly than the average	

Overview

Industrial machinery mechanics—often called *machinery maintenance mechanics* or *industrial machinery repairers*—inspect, maintain, repair, and adjust industrial production and processing machinery and equipment to ensure its proper operation in various industries.

History

Before 1750 and the beginning of the Industrial Revolution in Europe, almost all work was done by hand. Families grew their own food, wove their own cloth, and bought or traded very little. Gradually the economic landscape changed. Factories mass-produced products that had once been created by hand. The spinning jenny, a multiple-spindle machine for spinning wool or cotton, was one of the first machines of the Industrial Revolution. After it came a long procession of inventions and developments, including the steam engine, power loom, cotton gin, steamboat, locomotive, telegraph, and Bessemer converter. With these machines came the need for people who could maintain and repair them.

Mechanics learned that all machines are based on six configurations: the lever, the wheel and axle, the pulley, the inclined plane, the wedge, and the screw. By combining these elements in more complex ways, the machines could do more work in less time than people or animals could do. Thus, the role of machinery mechanics became vital in keeping production lines running and businesses profitable.

The Industrial Revolution continues even today, although now it is known as the Age of Automation. As machines become more numerous and more complex, the work of the industrial machinery mechanic becomes even more necessary.

The Job

The types of machinery on which industrial machinery mechanics work are as varied as the types of industries operating in the United States today. Mechanics are employed in metal stamping plants, printing plants, chemical and plastics plants—almost any type of large-scale industrial operation that can be imagined. The machinery in these plants must be maintained regularly. Breakdowns and delays with one machine can hinder a plant's entire operation, which is costly for the company.

Preventive maintenance is a major part of mechanics' jobs. They inspect the equipment, oil and grease moving components, and clean and repair parts. They also keep detailed maintenance records on the equipment they service. They often follow blueprints and engineering specifications to maintain and fix equipment.

When breakdowns occur, mechanics may partially or completely disassemble a machine to make the necessary repairs. They replace worn bearings, adjust clutches, and replace and repair defective parts. They may have to order replacement parts from the machinery's manufacturer. If no parts are available, they may have to make the necessary replacements, using milling machines, lathes, or other tooling equipment. After the machine is reassembled, they may have to make adjustments to its operational settings. They often work with the machine's regular operator to test it. When repairing electronically controlled machinery, mechanics may work closely with electronic repairers or electricians who maintain the machine's electronic parts.

Often these mechanics can identify potential breakdowns and fix problems before any real damage or delays occur. They may notice that a machine is vibrating, rattling, or squeaking, or they may see that the items produced by the machine are flawed. Many types of new machinery are built with programmed internal evaluation systems that check the accuracy and condition of equipment. This assists mechanics in their jobs, but it also makes them responsible for maintaining the check-up systems.

Machinery installations are becoming another facet of a mechanic's job. As plants retool and invest in new equipment, they rely on mechanics to properly situate and install the machinery. In many plants, millwrights traditionally did this job, but as employers increasingly seek workers with multiple skills, industrial machinery mechanics are taking on new responsibilities.

Industrial mechanics may use a wide range of tools when doing preventive maintenance or making repairs. For example, they may use simple tools such as a screwdriver and wrench to repair an engine or a hoist to lift a printing press off the ground. Sometimes they may have to solder or weld equipment. They use power and hand tools and precision measuring instruments. In some shops, mechanics troubleshoot for the entire plant's operations. Others may become experts in electronics, hydraulics, pneumatics, or other specialties.

Requirements

High School

While most employers prefer to hire those who have completed high school, opportunities do exist for those without a diploma as long as they have had some kind of related training. High school courses in mechanical drawing, general mathematics, algebra, geometry, physics, computers, and electronics are important.

Postsecondary Training

In the past, most industrial machinery mechanics learned the skills of the trade informally by spending several years as helpers in a particular factory. Currently, as machinery has become more complex, more formal training is necessary. Many mechanics are learning the trade through apprenticeship programs sponsored by a local trade union. Apprenticeship programs usually last four years and include both on-the-job and related classroom training. In addition to the use and care of machine and hand tools, apprentices learn the operation, lubrication, and adjustment of the machinery and equipment they will maintain. In class they learn shop mathematics, blueprint reading, safety, hydraulics, welding, and other subjects related to the trade.

Students may also obtain training through vocational or technical schools. Useful programs are those that offer machine shop courses and provide training in electronics and numerical control machine tools.

Other Requirements

Students interested in this field should possess mechanical aptitude and manual dexterity. Good physical condition and agility are necessary because mechanics sometimes have to lift heavy objects, crawl under large machines, or climb to reach equipment located high above the factory floor.

These workers are responsible for valuable equipment and are often called upon to exercise considerable independent judgment. Because of technological advances, they should be willing to learn the requirements of new machines and production techniques. When a plant purchases new equipment, the equipment's manufacturer often trains plant employees in proper operation and maintenance. Technological change requires mechanics to have adaptability and inquiring minds.

Exploring

Students interested in this field should take as many shop courses as they can. Exploring and repairing machinery such as automobiles and home appliances will sharpen the skills of those who are mechanically inclined. Students may also land part-time work or summer jobs in an industrial plant that give them opportunities to observe industrial repair work being done.

Employers

Industrial machinery mechanics work in a wide variety of plants and are employed in every part of the country, although employment is concentrated in industrialized areas. According to the U.S. Bureau of Labor Statistics, most mechanics work in industries such as food processing, textile mill products, chemicals, fabricated metal products, and primary metals. Others work for government agencies, public utilities, mining companies, and other facilities where industrial machinery is used.

Starting Out

Jobs can be obtained by directly applying to companies that use industrial equipment or machinery. The majority of mechanics work for manufacturing plants. These plants are found in a wide variety of industries, including the automotive, plastics, textile, electronics, packaging, food, beverage, and aerospace industries. Chances for job openings may be better at a large plant. New workers are generally assigned to work as helpers or trainees.

Prospective mechanics also may learn of job openings or apprenticeship programs through local unions. Industrial mechanics may be represented by one of several unions, depending on their industry and place of employment. These unions include the United Automobile, Aerospace and Agricultural Implement Workers of America; the United Steelworkers of America; the International Union of Electronic, Electrical, Salaried, Machine, and Furniture Workers; and the International Association of Machinists and Aerospace Workers. Private and state employment offices are other good sources of job openings.

Advancement

Those who begin as helpers or trainees usually become journey workers in four years. Although opportunities for advancement beyond this rank are somewhat limited, industrial machinery mechanics who learn more complicated machinery and equipment can advance into higher-paying positions. The most highly skilled mechanics may be promoted to master mechanics. Those who demonstrate good leadership and interpersonal skills can become supervisors. Skilled mechanics also have the option of becoming machinists, numerical control tool programmers, precision metalworkers, packaging machinery technicians, and robotics technicians. Some of these positions do require additional training, but the skills of a mechanic readily transfer to these areas.

Earnings

In 1997, the average annual earnings for industrial machinery mechanics was around $31,600, according to the U.S. Bureau of Labor Statistics. Apprentices generally earn lower wages and earn incremental raises as they advance in their training. Earnings vary based on experience, skills, type of industry, and geographic location. For example, mechanics employed in the textile industry generally earn wages at the low end of the scale, with workers in the automotive, metalworking, and aircraft industries earning wages at the high end. Mechanics in the Midwest typically earn higher salaries than those in the South. Those working in union plants generally earn more than those in nonunion plants. Most industrial machinery mechanics are provided with benefit packages, which can include paid holidays and vacations; medical, dental, and life insurance; and retirement plans.

Work Environment

Industrial machinery mechanics work in all types of manufacturing plants, which may be hot, noisy, and dirty or relatively quiet and clean. Mechanics frequently work with greasy, dirty equipment and need to be able to adapt to a variety of physical conditions. Because machinery is not always accessible, mechanics may have to work in stooped or cramped positions or on high ladders.

Although working around machinery poses some danger, with proper safety precautions, this risk is minimized. Modern machinery includes many safety features and devices, and most plants follow good safety practices. Mechanics often wear protective clothing and equipment, such as hard hats and safety belts, glasses, and shoes.

Mechanics work with little supervision and need to be able to work well with others. They need to be flexible and respond to changing priorities, which can result in interruptions that pull a mechanic off one job to repair a more urgent problem. Although the standard workweek is 40 hours, over-time is common. Because factories and other sites cannot afford breakdowns, industrial machinery mechanics may be called to the plant at night or on weekends for emergency repairs.

Outlook

The U.S. Bureau of Labor Statistics predicts that employment will grow more slowly than the average through 2006 for industrial machinery mechanics. Some industries will have a greater need for mechanics than others. Much of the new automated production equipment that firms are purchasing has its own self-diagnostic capabilities and is more reliable than older equipment. Although this machinery still needs to be maintained, most job openings will stem from the replacement of transferring or retiring workers.

Certain industries are extremely susceptible to changing economic factors and reduce production activities in slow periods. During these periods, companies may lay off workers or reduce hours. Mechanics are less likely to be laid off than other workers as machines need to be maintained regardless of production levels. Slower production periods and temporary shutdowns are often used to overhaul equipment. Nonetheless, employment opportunities are generally better at companies experiencing growth or stable levels of production.

Because machinery is becoming more complex and automated, mechanics need to be more highly skilled than in the past. Mechanics who stay up-to-date with new technologies, particularly those related to electronics and computers, will be best prepared to meet the needs of companies that use these workers.

For More Information

For information about scholarships, contact:

Association for Manufacturing Technology
7901 Westpark Drive
McLean, VA 22102-4269
Tel: 800-544-3597
Web: http://www.mfgtech.org

For information about apprentice programs, contact:

International Union, United Automobile, Aerospace and Agricultural Implement Workers of America
Skilled Trades Department
8000 East Jefferson Avenue
Detroit, MI 48214
Tel: 313-926-5000
Web: http://www.uaw.org

For information about the machining industry and careers in it, contact:

National Tooling & Machining Association
9300 Livingston Road
Fort Washington, MD 20744
Tel: 301-248-6200
Web: http://www.ntma.org

Job and Die Setters

Overview

Job and die setters, also known as *setup operators* or *setup workers,* prepare machine tools and production tools for others to use. They set up jigs, fixtures, cutting tools, and stamping tools on machines that are used for the shaping of metal. They also instruct machine tool operators on how to use the machines, and they make minor repairs and adjustments as needed during production.

History

Even in early times, tools were used to make things. The wood lathe actually dates back to the ancient world and was probably a variation of the potter's wheel. It works by mechanically rotating a workpiece against a stationary cutting tool. During the Industrial Revolution, as machine tools were invented and became more widely used, they became more complex and precise. During this time, lathes were adapted for cutting metal, and by the late 1700s, British inventor Henry Maudslay had devised the first screw-cutting lathe of high quality.

As more and more machine tools were developed, their use, in part, was responsible for the design of early mass-production methods in the United States. Setting up the tools and machines properly, so that they performed their functions exactly as intended, began to require more knowledge and skill. A need developed for specialized workers who understood the machining process to prepare and adjust these tools for operation. From this need came the job of the setup worker.

The latest developments began in the 1950s and 1960s with the introduction of numerically controlled machine tools. A numerical control system regulates machine performance by interpreting coded numerical data. Further improvements in machine tools have paralleled the advances in computer technology. Setup workers have had to stay current with all these changes. Today, with machine tools even more numerous, precise, and complex, manufacturing facilities and shops depend on these specialized setters to make certain that tools perform as intended.

The Job

Job and die setters prepare equipment for their co-workers, the machine tool operators, who run machine tools. Before setting up a tool, they study blueprints or other specifications to plan the sequence of machining operations and decide on the method for holding the workpiece in place.

They then set up the fixtures that hold the pieces being worked on and check the position of the workpiece to make sure it is correct. They select and set up cutting tools to shave off the exact amount of material. They may also set up planers, milling machines, grinders, presses, turret lathes, and automatic machine centers. After positioning the tool, setup workers move controls to position the tool and workpiece in the correct relation to each other. They then set the speed, feed, and depth of the cut. To ensure that their setup is correct, they may use measuring tools, such as micrometers or gauges.

After set up, they often test the machine by running off several pieces to make certain that their settings are accurate. They then turn the job over to the machine tool operator. Setup workers show or tell operators how to run the machines. They warn them about potential difficulties and explain how to avoid them. During a job run, setters make adjustments when necessary, and they change cutting tools and adjust specifications if needed.

Setters may work on only one type of machine or on most of the machines in a plant or factory; they may work with only one metal, alloy, or plastic, or with several; and they may work on only one product, or on a variety. Specialists include buffing-line set up workers; thread tool grinder setup operators; trim-machine adjusters; slitter service and setters; honing workers; spline-rolling machine operators; job setters; and punch-press, spring coiling machine, threading-machine, grinder machine, and molding-and-coremaking machine workers.

Requirements

High School

Employers usually prefer to hire high school graduates who have good basic skills and have taken algebra, geometry, and trigonometry. Metal or machine shop and blueprint reading are also helpful. Competency in English is necessary because of the operating explanations a setup operator must give to machine tenders.

Postsecondary Training

Job and die setters begin their careers as machine tool operators. After they become skilled and knowledgeable about machining processes and the products being manufactured, they may be promoted to setup work. It often takes several years of observing and assisting experienced co-workers to become competent in adjusting the feed and speed of machines and changing cutting tools. Setup operators also learn how to read blueprints and plan work sequence in addition to making machine adjustments. Sometimes this training occurs in a formal training program that includes classroom instruction as well as on-the-job observation.

Other Requirements

To be a successful setup worker, an interest in machines and an aptitude for mechanical processes are necessary. The ability to work independently, with precision and attention to detail, is important. Finally, strong communications skills are a plus, since setters are responsible for showing machine tool operators how to carry out machining operations.

Exploring

Machine shop courses in high school or technical school provide one way of gaining some experience in this work. You might join a student organization, such as Vocational Industrial Clubs of America (VICA) or the Technology Student Association (TSA), if it is active at your school. Summer jobs as unskilled workers are sometimes available in manufacturing plants that have large machine shops. Interacting with the skilled workers in such a plant can provide you with a beginning knowledge of machine tools and the work of setup operators.

Employers

Most setup operators work in factories that make cars, trucks, planes, and other types of machinery. While they work in every state, most jobs are located in the midwestern, northeastern, and far west sections of the country. Most are found in or near major metropolitan areas, such as Los Angeles, Chicago, New York, Philadelphia, Cleveland, and Detroit.

Starting Out

Job seekers should apply directly to machine shops or factories. Workers will begin as machine tool operators to gain skill and experience. After several years as a machine operator, and if enough proficiency has been shown in this entry-level position, the next step is to apply for a job setter training position.

Advancement

Advancing from setter to shop supervisor is considered a natural step by most employees. Some also further their careers by taking classes at a technical school or a two-year college. This may help them advance to more skilled work, such as numerical control tool programming or tool designing, which is more lucrative than operating machines. Skilled setup operators might also consider opening their own shops. To be successful, however, workers should prepare themselves by acquiring business skills.

Earnings

According to the Bureau of Labor Statistics, job and die setters earned an average annual wage of about $31,000. Experienced job and die setters earn over $40,000. Many jobs in machining require overtime, and workers are eligible for overtime pay. Almost all setup operators receive a benefits package that includes paid holidays and vacations, health insurance, and retirement plans.

Work Environment

Overall working conditions for setters are good. Machine shops are not too noisy and are usually well lit and ventilated, and air-conditioned during the summer months. Some places, such as forge shops, may be noisy, however, and many job and die setters work on production lines in factories.

Setters spend much of their workdays on their feet. Although cranes do most of the heavy lifting, die setters do work with heavy chains and use large wrenches to tighten the dies in place. Working around high-speed machines can be dangerous, and safety measures, including machine guards and safety glasses, and rules are strictly enforced.

Outlook

Overall employment of job and die setters is expected to decline in the next several years. In general, employment is affected by the rate of technological improvements being used in industry, the demand for the type of goods produced, the effects of international trade, and the reorganization of production processes. Computer-controlled equipment allows workers to simultaneously operate a greater number of machines and makes setup easier. This reduces the time spend on setting up each machine and the number of workers needed. However, if setters keep their skills current or develop a combination of skills, they will continue to be in demand as other workers transfer or leave the labor force.

For More Information

For information on apprenticeships in the machine tools trades, contact:

International Union, United Automobile, Aerospace, and Agricultural Implement Workers of America
Skilled Trades Department
8000 East Jefferson
Detroit, MI 48214
Tel: 313-926-5000
Web: http://uaw.org

International Union of Electronic, Electrical, Salaried, Machine, and Furniture Workers
1126 16th Street, NW
Washington, DC 20036
Tel: 202-785-7200
Web: http://www.iue.org

For literature on training and careers in the machine tools trades, contact:

National Tooling & Machining Association
9300 Livingston Road
Fort Washington, MD 20744
Tel: 301-248-6200
Web: http://www.ntma.org

Tooling and Manufacturing Association
1177 South Dee Road
Park Ridge, IL 60068
Tel: 847-825-1120
Web: http://www.tmanet.org

Machine Tool Operators

School Subjects	Mathematics Technical/Shop
Personal Skills	Mechanical/manipulative Technical/scientific
Work Environment	Primarily indoors Primarily one location
Minimum Education Level	High school diploma
Salary Range	$17,000 to $26,000 to $40,000
Certification or Licensing	None available
Outlook	Decline

Overview

Machine tool operators operate or tend one or more types of machine tools that have already been set up for a job. These tools cut, drill, grind, bore, mill or use a combination of methods to cut or finish pieces of metal or plastic products. These machines include lathes, boring mills, drilling and screw machines, jig grinders and borers, and milling machines. Some machine tool operators work with numerically controlled (NC) equipment.

History

At one time, before modern manufacturing procedures, goods were made individually by one craftsworker. As shops grew larger and employed more workers, the process changed. The steps involved in creating a product were separated into a series of easy tasks, which workers could learn quickly and do repetitively. Each worker was responsible for one part of the process.

Various tools were used in the manufacturing process, even in early times. One of the earliest machine tools, the wood lathe, actually dates back to ancient times and is probably a variation of the potter's wheel. It performs by mechanically rotating a workpiece against a stationary cutting tool.

With the advent of the Industrial Revolution, machine tools became more advanced and more widely used. During this time, lathes were adapted for cutting metal, and by the late 1700s, British inventor Henry Maudslay had devised the first screw-cutting lathe of high quality. By 1775, British industrialist John Wilkinson invented a boring machine that made holes in metal with precise accuracy. The planer—a metal-cutting device that holds a workpiece in place while a cutting tool moves back and forth—was developed 25 years later. The planing tool allowed holes and flat surfaces to be smoothed to necessary degrees.

Technological improvements in machine tools impacted the burgeoning Industrial Revolution. The use of such tools was, in fact, responsible in part for the design of early mass-production methods in the United States. However, the most rapid spurt in the development of machine tools has come since World Wars I and II. During the wars, it was necessary to build tanks, planes, jeeps, ships, and guns rapidly and accurately, so machines had to be devised that would turn out the thousands of pieces required.

The latest developments began in the 1950s and 1960s with the design of numerically controlled machine tools. A numerical control system regulates the performance of a machine tool by interpreting coded numerical data which then directs the positioning and actual machining of the tool. Further improvements in machine tools have paralleled the advances in computer technology. Methods from which machine tooling has benefited include computer-integrated manufacturing (CIM), computer-aided design (CAD), and robotics.

The Job

Machine tool operators tend to the operation of one or two machines that have already been set up by a job setter or setup operator. Although some workers are known by the specific machines that they are responsible for (e.g., lathe operator, drilling machine operator), most are trained to work on a variety of machines.

A typical machine tool operator, for example, may tend a drilling machine. The operator starts the drill, inserts a piece of metal stock into the guide that holds it during machining, pulls down the lever of the drill press until the piece is drilled the prescribed distance, and releases the lever. Completed parts are then removed from the machine and placed in a bin.

During the machining process, the operator watches to make sure that the machine is working properly. When needed, the operator adds coolants and lubricants to the machinery and the workpiece. Except during breakdowns or while new stock is being brought up for machining, the machine tool operator generally repeats the same process until the batch of pieces is completed.

In some shops though, an operator tends a series of the machines that shape and finish a machine part, and may even do some programming. Skill requirements vary from job to job. When a new program is loaded, it often must be adjusted to obtain the desired results. A machinist or tool programmer usually performs this function.

Because numerical control (NC) machine tools are expensive, operators who work on these more advanced machines must carefully monitor operations to prevent costly damage to cutting tools or other parts. The extent to which this is required, however, depends on the type of job and equipment being used. In some cases, the operator may only need to watch a machine as it functions, and therefore can set up and operate more than one machine at a time. Other jobs may require frequent loading and unloading, tool changing, or programming. Operators may check finished parts with micrometers, gauges, or other precision inspection equipment to ensure they meet specifications, although NC machine tools are increasingly able to do this as parts are produced.

Requirements

High School

A high school diploma is preferred by most employers of machine tool operators. Classes in algebra, geometry, and drafting or mechanical drawing are excellent preparation for this job. Machine shop also is very helpful.

Postsecondary Training

Most workers in this occupation learn their skills on the job. Trainees begin by observing and helping experienced workers, sometimes in formal training programs. As part of their training, they advance to more difficult and complex tasks, such as adjusting feed speeds or changing cutting tools. They also learn to check gauges and do basic shop calculations. Eventually, they become responsible for their own machines.

Other Requirements

Students interested in this work should have better than average mechanical aptitude, manual dexterity, and an interest in machines. The ability to pay close attention to the task, even when it becomes repetitive and dull, is essential. As machinery becomes more complex and shop floor organization changes, employers are increasingly looking for people with good communications skills. Finally, because machine tool operators spend most of their days standing at machines, a certain amount of physical stamina is required.

Exploring

Hobbies such as building models and working with wood and other materials will give you practical experience in fundamental machining concepts. If you want to explore the occupation further, high school or vocational school shop classes teach technical theory and machining techniques.

There are several relevant student organizations such as Vocational Industrial Clubs of America (VICA) and the Technology Student Association (TSA). If you wish to see machine tool operators in action, you might ask teachers or guidance counselors to set up a visit to a local plant, or you might investigate a summer or part-time job as a general helper in a nearby shop.

Employers

According to the U.S. Bureau of Labor Statistics, most machine tool operators are employed in five manufacturing industries: fabricated metal products, industrial machinery and equipment, miscellaneous plastic products, transportation equipment, and primary metals. Shops and plants are most often found in the industrial areas of the Northeast and Midwest as well as California.

Starting Out

Job seekers should apply directly to the personnel offices of machine shops and factories. Entry-level workers may start out by doing a wide variety of jobs at the plant, first learning skills by observing experienced workers, and later by working under supervision until they are capable of working independently. This type of training usually lasts about one to two years. Job openings for machine tool operators may be listed in classified ads of newspapers as well as with state and private employment agencies.

Advancement

Becoming a professional, skilled operator with commensurate wages often takes a number of years. In addition, it is generally only after several years' experience that a machine tool operator can advance to the position of setup operator. An operator who can read blueprints and use measurement tools, and is willing to try new methods, is more likely to be moved into a supervisory job or to a more versatile position such as a numerical control programmer or a job setter. Operators may also transfer to training programs for other related occupations, such as precision machinist or toolmaker.

Earnings

The annual wage for machine tool operators varies according to which machines they run, the size of the facility, and where it is located. According to the U.S. Bureau of Labor Statistics, the average annual wage ranges from $22,000 to almost $27,000. Many workers often work more than 40 hours a week and earn overtime pay. Earnings also vary considerably by industry. Operators who work in the manufacturing of transportation equipment earn substantially more than those who work in rubber and plastic products manufacturing.

Many machine tool firms have traditional benefit plans, including retirement programs to which both the employer and the employee contribute. Most operators are also eligible for paid vacations, sick leave, and group hospitalization insurance.

In many manufacturing operations, the plant closes to change over machinery in order to make new models. Workers are seldom paid for this downtime.

Work Environment

In general, the workweek is 40 hours, but often, when large orders have to be met quickly, employees are asked to work late and on Saturdays. Work is performed exclusively indoors. Conditions can be somewhat dangerous, particularly because of the high speeds and pressures at which these machines operate. Therefore, protective equipment must be used, and safety rules must be observed. There are other minor hazards as well, including, for example, skin irritations from coolants used on cutting and drilling machines. Operators must wear goggles and avoid wearing loose clothing that could get caught in machinery. Most machine shops are clean, and well lit and ventilated, although some older ones may be less so. The shops are often noisy because of the operating machinery.

The main drawback to this profession is the repetitive and sometimes boring nature of the work. Machine tool operators typically spend hours each day performing the same task, over and over. However, the fairly high wage for work that is relatively easy to learn and perform may compensate for the tedium of the job.

Outlook

Overall employment of machine tool operators is expected to decline in the coming years. The main reason is the change to labor-saving machinery. In order to remain competitive, many firms have adopted new technologies such as computer-controlled machine tools to improve quality and lower production costs. Computer-controlled equipment allows operators to tend a greater number of machines simultaneously, and thereby reduces the number of employees needed. However, employment of operators who are skilled in the use of these machines is expected to increase, while positions for manual machine operators continue to decline.

Also, the demand for machine tool operators parallels the demand for the products they produce. In recent years, plastic has been substituted for metal in many manufacturing parts. If this trend continues, the demand for machine tool operators in plastics manufacturing will be greater than for those in metals.

Even with the decline in new positions, there should still be many job possibilities for machine operators. It is estimated that, within the next 10 to 15 years, 60 percent of the existing workforce will be leaving the occupation and will have to be replaced.

For More Information

For information on training and jobs for machine tool operators, contact:

International Union, United Automobile, Aerospace, and Agricultural Implement Workers of America
Skilled Trades Department
8000 East Jefferson Avenue
Detroit, MI 48214
Tel: 313-926-5000
Web: http://www.uaw.org

International Union of Electronic, Electrical, Salaried, Machine, and Furniture Workers
1126 16th Street, NW
Washington, DC 20036
Tel: 202-785-7200
Web: http://www.iue.org

For literature on careers in the machine tool industry, contact:

National Tooling & Machining Association
9300 Livingston Road
Fort Washington, MD 20744
Tel: 301-248-6200
Web: http://www.ntma.org

For literature on careers and training in the machine tools trades, contact:

Precision Machined Products Association
6700 West Snowville Road
Brecksville, OH 44141-3292
Tel: 216-526-0300
Web: http://www.pmpa.org

Tooling and Manufacturing Association
1177 South Dee Road
Park Ridge, IL 60068
Tel: 847-825-1120
Web: http://www.tmanet.com

Manufacturing Supervisors

Mathematics Technical/Shop	School Subjects
Mechanical/manipulative Technical/scientific	Personal Skills
Primarily indoors Primarily one location	Work Environment
High school diploma	Minimum Education Level
$18,200 to $33,300 to $51,000	Salary Range
None available	Certification or Licensing
Decline	Outlook

Overview

Manufacturing supervisors are bosses. They monitor employees' working conditions and effectiveness. They ensure that work is being done correctly and on time. To carry out this work, supervisors maintain their employees' work schedules, train new workers, and issue warnings to employees who do not fulfill their duties properly or who violate certain established rules.

History

Manufacturing has undergone many technological developments, from innovations in fuel-powered machinery to the assembly line. As these processes became more complex, no one individual worker could be responsible for the production of one particular item. The production of an individual item was part of a long process and involved passing through many workers' hands. If one worker caused a defect in the product, it was not always easy to find the worker who caused the defect and correct the problem. The role of the supervisor emerged as a means of keeping track of the work of numerous employees involved in the production process, which allowed them to locate problems and keep production running smoothly.

The Job

The primary roles of manufacturing supervisors are to oversee their employees and ensure the effectiveness of the production process. They are responsible for the amount of work and the quality of work being done by the employees under their direction. Supervisors make work schedules, keep production and employee records, and plan on-the-job activities. Their work is highly interpersonal. They not only monitor employees, but also guide workers in their efforts and are responsible for disciplining and counseling poor performers as well as recommending good performers for raises and promotions. They also make sure that safety regulations and other rules and procedures are being followed.

Manufacturing supervisors may work in small companies, such as custom-built furniture shops, or large industrial factories where cars or food products are made. Supervisors answer to company managers, who in turn direct supervisors and make sure they are doing their jobs properly. Another important part of the supervisor's job is to act as a link between factory workers and company managers who are in charge of production. Supervisors announce new company policies and plans to workers and report to their managers about the success of their employees, any problems they may be having, or other issues pertaining to the factory workers they monitor. Supervisors also may meet with other company supervisors to discuss progress toward company goals, company operations, and employee performance. In companies whose employees belong to labor unions, supervisors must know and follow all work-related guidelines outlined by labor-management contracts.

Requirements

High School

High school students interested in becoming manufacturing supervisors should take high school courses in business, math, and science. A high school diploma is the minimum requirement for supervisory positions. Many supervisors begin as factory workers and are promoted to supervisors after demonstrating their ability on the job.

One or two years of college or technical school may also be valuable to employers, depending on the type of employer and the type of manufacturing company. College courses in business, business administration, industrial relations, math, and science are important for familiarizing prospective supervisors with the type of knowledge they will need for the tasks they will encounter.

Manufacturing supervisors deal with many people on a highly personal level. They must direct, guide, and discipline others, so strong leadership qualities are important characteristics for supervisors. Good communications skills, and the ability to motivate people and maintain morale are also qualities of successful supervisors.

Starting Out

Many supervisors enter their jobs by moving up from factory worker positions. They may also apply for supervisory positions from outside the company. Companies that hire manufacturing supervisors look for experience, knowledge of the job or industry, organizational skills, and leadership abilities. Supervisory positions may be found in classified ads, but for those just looking to get started, part-time or full-time jobs in a factory setting may help provide some experience and familiarization with what the work of supervisors entails. It can lead to the experience necessary for obtaining a position as a supervisor.

Advancement

In most manufacturing companies, a business degree along with in-house training are required for moving up into higher positions. From the position of supervisor, one may advance to manager or head of an entire department.

Earnings

Supervisors made an average of about $33,280 in 1996. The middle 50 percent earned between $24,960 and $42,640 while the lowest paid supervisors earned less than $18,200. The highest percent earned over $51,000. They also receive health benefits in addition to their salaries and overtime pay.

Work Environment

Most supervisors work on the manufacturing or factory floor. They may be on their feet most of the time, which can be tiring, and work near loud and hazardous machines. Supervisors may begin their day early so that they arrive before their workers, and they may stay later than their workers. Some may work for plants that operate around the clock, so they may work late shifts and weekend hours and holidays. Sometimes the best hours go to the supervisors who have been with the company the longest. A lot of plant downsizing and restructuring has led to fewer supervisors. This means supervisors have more employees under their supervision and sometimes work longer hours and must hold more responsibilities.

Outlook

To some extent the future of the manufacturing supervisor job depends on the individual industry. However, in manufacturing as a whole, employment of supervisors is expected to decline because each supervisor will be overseeing more workers. Corporate downsizing and increasing use of computers for some supervisory responsibilities mean fewer supervisors will be needed.

For More Information

American Management Association
135 West 50th Street
New York, NY 10020-1201
Tel: 212-586-8100

National Association of Manufacturers
1331 Pennsylvania Avenue, NW
Washington, DC 20004
Tel: 202-637-3000

Mechanical Engineering Technicians

English Mathematics Physics	School Subjects
Mechanical/manipulative Technical/scientific	Personal Skills
Primarily indoors Primarily one location	Work Environment
Associate's degree	Minimum Education Level
$23,000 to $39,000 to $64,000	Salary Range
Voluntary	Certification or Licensing
About as fast as the average	Outlook

Overview

Mechanical engineering technicians work under the direction of mechanical engineers to design, build, maintain, and modify many kinds of machines, mechanical devices, and tools. They work in a wide range of industries and in a variety of specific jobs within every industry.

Mechanical engineering technicians review mechanical drawings and project instructions, analyze design plans to determine costs and practical value, sketch rough layouts of proposed machines or parts, assemble new or modified devices or components, test completed assemblies or components, analyze test results, and write reports.

History

Mechanical engineering dates back to ancient times, when it was used almost exclusively for military purposes. Perhaps the Romans were the first to use the science for nonmilitary projects, such as aqueducts, roads, and bridges, although many if not most of these structures were built to advance military objectives.

With the advent of the Industrial Revolution and the use of machines for manufacturing, mechanical engineering technology took a giant step forward. One of the most important figures in this revolution was Eli Whitney. Having received a government contract in 1798 to produce 15,000 muskets, he hired not gunsmiths, but mechanics. At that time, all articles, including muskets, were built one by one by individual craft workers. No two muskets were ever alike.

Whitney took a different approach. For two years after receiving the contract, he focused on developing and building special-purpose machines, and then trained mechanics to make specific parts of the gun. When he was finished, Whitney had invented new machine tools and attachments, such as the milling machine and jig; made real the concept of interchangeable parts; and paved the way for the modern manufacturing assembly line.

This manufacturing process required not only ingenious inventors and skilled mechanics to operate the machines, but also skilled assistants to help develop new machines, set or reset tolerances, maintain and repair operational equipment, and direct, supervise, and instruct workers. These assistants are today's mechanical engineering technicians, a crucial part of today's engineering team. In addition to manufacturing, they are employed in almost every application that uses mechanical principles.

The Job

Mechanical engineering technicians are employed in a broad range of industries. Technicians may specialize in any one of many areas including biomedical equipment, measurement and control, products manufacturing, solar energy, turbo machinery, energy resource technology, and engineering materials and technology.

Within each application, there are various aspects of the work with which the technician may be involved. One phase is research and development. In this area, the mechanical technician may assist an engineer or scientist in the design and development of anything from a ballpoint to a sophisticated measuring device. These technicians prepare rough sketches and layouts of the project being developed.

In the design of an automobile engine, for example, engineering technicians make drawings that detail the fans, pistons, connecting rods, and flywheels to be used in the engine. They estimate cost and operational qualities of each part, taking into account friction, stress, strain, and vibration. By performing these tasks, they free the engineer to accomplish other research assignments.

A second common type of work for mechanical engineering technicians is testing. For products such as engines, motors, or other moving devices, technicians may set up prototypes of the equipment to be tested and run performance tests. Some tests require one procedure to be done repeatedly, while others require that equipment be run over long periods of time to observe any changes in operation. Technicians collect and compile all necessary data from the testing procedures and prepare reports for the engineer or scientist.

In the manufacture of a product, preparations must be made for its production. In this effort, mechanical engineering technicians assist in the product design by making final design layouts and detailed drawings of parts to be manufactured and of any special manufacturing equipment needed. Some test and inspect machines and equipment or work with engineers to eliminate production problems.

Other mechanical engineering technicians examine plans and drawings to determine what materials are needed and prepare lists of these materials, specifying quality, size, and strength. They also may estimate labor costs, equipment life, and plant space needed. After the product is manufactured, some mechanical engineering technicians may help solve storage and shipping problems, while others assist in customer relations when product installation is required.

Some engineering technicians work with tool designers. They help in preparing sketches of designs for cutting tools, jigs, special fixtures, and other devices used in mass production. Frequently, they redesign existing tools to improve their efficiency.

Requirements

High School

Preparation for this career begins in high school. Although entrance requirements to associate programs vary somewhat from school to school, mathematics and physical science form the backbone of a good preparatory curriculum. Classes should include algebra, geometry, science, computer science, mechanical drawing, shop, and communications.

Postsecondary Training

Associate's degree or two-year mechanical technician programs are designed to prepare students for entry-level positions. Most programs accredited by the Accreditation Board for Engineering and Technology (ABET) offer one year of a basic program with a chance to specialize in the second year. The first year of the program generally consists of courses in college algebra and trigonometry, science, and communications skills. Other classes introduce students to the manufacturing processes, drafting, and language of the industry.

The second year's courses focus on mechanical technology. These include fluid mechanics, thermodynamics, tool and machine design, instruments and controls, production technology, electricity, and electronics. Many schools allow their students to choose a major in the second year of the program, which provides training for a specific area of work in the manufacturing industry.

Certification or Licensing

Many mechanical engineering technicians choose to become certified by the National Institute for Certification in Engineering Technologies (NICET). To become certified, a technician must combine a specific amount of job-related experience with a written examination. Certifications are offered at several levels of expertise. Such certification is generally voluntary, although obtaining certification shows a high level of commitment and dedication that employers find highly desirable.

Mechanical engineering technicians are encouraged to become affiliated with professional groups, such as the American Society of Certified Engineering Technicians, that offer continuing education sessions for members. Some mechanical engineering technicians may be required to belong to unions.

Other Requirements

Technicians need mathematical and mechanical aptitude. They understand abstract concepts and apply scientific principles to problems in the shop or laboratory, in both the design and the manufacturing process. They are interested in people and machines and have the ability to carry out detailed work. They analyze sketches and drawings and possess patience, perseverance, and resourcefulness. Additionally, they have good communication skills and can present both spoken and written reports.

Exploring

You may be able to obtain part-time or summer work in a machine shop or factory. This type of work usually consists of sweeping floors and clearing out machine tools, but it offers an opportunity to view the field firsthand and also demonstrates interest to future employers. Field trips to industrial laboratories, drafting studios, or manufacturing facilities can offer overall views of this type of work. Hobbies like automobile repair, model making, and electronic kit assembling can also be helpful. Finally, any high school student interested in the engineering field should consider joining the Junior Engineering Technical Society (JETS).

Employers

Many engineering technicians work in durable goods manufacturing, primarily making electrical and electronic machinery and equipment, industrial machinery and equipment, instruments and related products, and transportation equipment. A sizable percentage work is in service industries, mostly in engineering and business services companies that do contract work for government, manufacturing, and other organizations.

The federal government employs engineering technicians in the Departments of Defense, Transportation, Agriculture, and Interior as well as the Tennessee Valley Authority and the National Aeronautics and Space

Administration. State and municipal governments also have engineering technicians working for them.

Starting Out

Schools offering associate's degrees in mechanical engineering technology and two-year technician programs usually help graduates find employment. At most colleges, in fact, company recruiters interview prospective graduates during their final semester of school. As a result, many students receive job offers before graduation. Other graduates may prefer to apply directly to employers, use newspaper classified ads, or apply through public or private employment services.

Advancement

As mechanical engineering technicians remain with a company, they become more valuable to the employer. Opportunities for advancement are available to those who are willing to accept greater responsibilities either by specializing in a specific field, taking on more technically complex assignments, or by assuming supervisory duties. Some technicians advance by moving into technical sales or customer relations. Mechanical technicians who further their education may choose to become tool designers or mechanical engineers.

Earnings

Salaries for mechanical engineering technicians vary depending on the nature and location of the job, employer, amount of training the technician has received, and number of years of experience.

In general, mechanical engineering technicians who develop and test machinery and equipment under the direction of an engineering staff earn between $28,000 and $50,000 a year. The average in 1997 was about $39,000, according to the U.S. Bureau of Labor Statistics. Some mechanical engineering technicians, especially those at the beginning of their careers, may make around $23,000 a year or less, while some senior technicians with special skills and experience may make from $50,000 to $64,000 a year or more.

These salaries are based upon the standard workweek. Overtime or premium time pay may be earned for work beyond regular daytime hours or workweek. Other benefits, depending on the company and union agreements, include paid vacation days, insurance, retirement plans, profit sharing, and tuition-reimbursement plans.

Work Environment

Mechanical engineering technicians work in a variety of conditions, depending on their field of specialization. Technicians who specialize in design may find that they spend most of their time at the drafting board or computer. Those who specialize in manufacturing may spend some time at a desk, but also spend considerable time in manufacturing areas or shops.

Conditions also vary according to industry. Some industries require technicians to work in foundries, die-casting rooms, machine shops, assembly areas, or punch-press areas. Most of these areas, however, are well lighted, heated, and ventilated. Moreover, most industries employing mechanical engineering technicians have strong safety programs.

Mechanical engineering technicians are often called upon to exercise decision-making skills, to be responsible for valuable equipment, and to act as effective leaders. At other times they carry out routine, uncomplicated tasks. Similarly, in some cases, they may coordinate the activities of others, while at other times, they are the ones supervised. They must be able to respond well to both types of demands. In return for this flexibility and versatility, mechanical engineering technicians are usually highly respected by their employers and coworkers.

Outlook

Job opportunities for mechanical engineering technicians are expected to grow as quickly as the average through the year 2006. Manufacturing companies will be looking for more ways to apply the advances in mechanical technology to their operations. Opportunities will be best for technicians who are skilled in new manufacturing concepts, materials, and designs. Many job openings also will be created by people retiring or leaving the field.

However, the employment outlook for engineering technicians is influenced by the economy. Hiring will fluctuate with the ups and downs of the nation's overall economic situation. Technicians whose jobs are defense-related may experience fewer opportunities because of recent defense cutbacks.

For More Information

For information about membership in a professional society for engineering technicians, contact:

American Society of Certified Engineering Technicians (ASCET)
PO Box 1348
Flowery Branch, GA 30542
Tel: 770-967-9173

For information about the field of mechanical engineering, contact:
American Society of Mechanical Engineers
Three Park Avenue
New York, NY 10016
Tel: 212-591-7000
Web: http://www.asme.org

For information on high school programs that provide opportunities to learn about engineering technology, contact:

Junior Engineering Technical Society, Inc.
1420 King Street, Suite 405
Alexandria, VA 22314-2794
Tel: 703-548-5387
Web: http://www.jets.org

For information on certification of mechanical engineering technicians, contact:

National Institute for Certification in Engineering Technologies
1420 King Street
Alexandria, VA 22314-2715
Tel: 888-476-4238
Web: http://www.nicet.org

Mechanical Engineers

School Subjects
Computer science
English
Mathematics

Personal Skills
Leadership/management
Technical/scientific

Work Environment
Primarily indoors
One location with some travel

Minimum Education Level
Bachelor's degree

Salary Range
$34,000 to $52,000 to $85,000

Certification or Licensing
Voluntary

Outlook
About as fast as the average

Overview

Mechanical engineers plan and design tools, engines, machines, and other mechanical systems that produce, transmit, or use power. Their work varies by industry and function. They may work in design, instrumentation, testing, robotics, transportation, or bioengineering, among other areas. The broadest of all engineering disciplines, it extends across many interdependent specialties. Mechanical engineers may work in production operations, maintenance, or technical sales, and many are administrators or managers.

History

In a general sense, mechanical engineering has existed for thousands of years. Pyramid building in ancient Egypt, for example, required extensive knowledge of engineering principles. Large blocks of two- and three-ton stone were quarried, transported, and positioned according to sophisticated designs.

Ancient Greeks and Romans were also great builders, but, unlike the Egyptians, they developed and made use of elaborate mechanical devices, including water pumps, machines for cutting screws, and treadmills that produced power for lifting heavy objects. Remarkably, the Greeks even invented a steam engine, but viewed it only as a curiosity or toy.

The term "engineer" was coined around the 14th century and applied to people who designed equipment for war. Their achievements were so important that the strength of a country's military became increasingly dependent upon their inventions. As these individuals applied their knowledge to civilian needs, new occupational terms developed. Engineers who worked on civilian projects came to be known as civil, as opposed to military, engineers. Later, engineers who concentrated on machinery and the generation of power were called mechanical engineers.

The modern field of mechanical engineering took root during the Renaissance. In this period, engineers focused their energies on developing more efficient ways to perform such ordinary tasks as grinding grain and pumping water. Water wheels and windmills were common energy producers at that time. Leonardo da Vinci, who attempted to design such complex machines as a submarine and a helicopter, best personified the burgeoning mechanical inventiveness of the period. One of the Renaissance's most significant inventions was the mechanical clock, powered first by falling weights and later by compressed springs.

Despite these developments, it was not until the Industrial Revolution that mechanical engineering took on its modern form. The steam engine, an efficient power producer, was introduced in 1712 by Thomas Newcomen to pump water from English mines. More than a half century later, James Watt modified Newcomen's engine to power industrial machines. In 1876, German Nicolaus Otto developed the internal combustion engine, which became one of the century's most important inventions. In 1847, a group of British engineers, who specialized in steam engines and machine tools, organized the Institution of Mechanical Engineers. The American Society of Mechanical Engineers was formed by 1880.

Mechanical engineering has rapidly expanded in the 20th century. More than 200,000 mechanical engineers are employed in the United States alone. Mass production systems allow large quantities of standardized goods to be made at a low cost, and mechanical engineers play a pivotal role in the design of these systems. In the second half of the 20th century, computers revolutionized production. Mechanical engineers now design mechanical systems on computers, and they are used to test, monitor, and analyze mechanical systems and factory production. Mechanical engineers realize this trend is here to stay.

The Job

The work of mechanical engineering begins with research and development. A company may need to develop a more fuel-efficient automobile engine, for example, or a cooling system for air-conditioning and refrigeration that does not harm the earth's atmosphere. A *research engineer* explores the project's theoretical, mechanical, and material problems. The engineer may perform experiments to gather necessary data and acquire new knowledge. Often, an experimental device or system is developed.

The *design engineer* takes information gained from research and development and uses it to plan a commercially useful product. To prevent rotting in a grain storage system, for example, a design engineer might use research on a new method of circulating air through grain. The engineer would be responsible for specifying every detail of the machine or mechanical system. Since the introduction of sophisticated software programs, mechanical engineers have increasingly used computers in the design process.

After the product has been designed and a prototype developed, the product is analyzed by *testing engineers*. A tractor transmission, for example, would need to be tested for temperature, vibration, dust, and performance under the required loads, as well as for any government safety regulations. If dust is penetrating a bearing, the testing engineer would refer the problem to the design engineer, who would then make an adjustment to the design of the transmission. Design and testing engineers continue to work together until the product meets the necessary criteria.

Once the final design is set, it is the job of the *manufacturing engineer* to come up with the most time- and cost-efficient way of making the product, without sacrificing quality. The amount of factory floor space, the type of manufacturing equipment and machinery, and the cost of labor and materials are some of the factors that must be considered. Engineers select the necessary equipment and machines and oversee their arrangement and safe operation. Other engineering specialists, such as *chemical, electrical,* and *industrial engineers,* may provide assistance.

Some types of mechanical systems—from machinery on a factory floor to a nuclear power plant—are so sophisticated that mechanical engineers are needed for operation and ongoing maintenance. With the help of computers, maintenance and operations engineers use their specialized knowledge to monitor these complex systems and to make necessary adjustments.

Mechanical engineers also work in marketing, sales, and administration. Because of their training in mechanical engineering, sales engineers can give customers a detailed explanation of how a machine or system works. They may also be able to alter its design to meet a customer's needs.

In a small company, a mechanical engineer may need to perform many, if not most, of the above responsibilities. Some tasks might be assigned to consulting engineers, who are either self-employed or work for a consulting firm.

Other mechanical engineers may work in a number of specialized areas. *Energy specialists* work with power production machines to supply clean and efficient energy to individuals and industries. *Application engineers* specialize in computer-aided design (CAD) systems. *Process engineers* work in environmental sciences to reduce air pollution levels without sacrificing essential services such as those provided by power stations or utility companies.

Requirements

High School

If you are interested in mechanical engineering as a career, you need to take courses in geometry, trigonometry, and calculus. Physics and chemistry courses are also recommended, as is mechanical drawing or computer-aided design, if they are offered at your high school. Communications skills are important for mechanical engineers because they interact with a variety of co-workers and vendors and are often required to prepare and/or present reports. English and speech classes would be helpful. Finally, because computers are such an important part of engineering, computer science courses are good choices.

Postsecondary Training

A bachelor's degree in mechanical engineering is usually the minimum educational requirement for entering this field. A master's degree, or even a Ph.D., may be necessary for obtaining some positions, such as those in research, teaching, and administration.

In the United States, there are more than 200 colleges and universities where engineering programs have been approved by the Accreditation Board for Engineering and Technology (ABET). Most of these institutions offer programs in mechanical engineering. Although admissions requirements vary slightly from school to school, most require a solid background in mathematics and science.

In a four-year undergraduate program, students typically begin by studying mathematics and science subjects, such as calculus, differential equations, physics, and chemistry. Course work in liberal arts and elementary mechanical engineering is also taken. By the third year, students begin to study the technical core subjects of mechanical engineering—mechanics, thermodynamics, fluid mechanics, design manufacturing, and heat transfer—as well as such specialized topics as power generation and transmission, CAD, and the properties of materials.

At some schools, there is a five- or six-year program that combines classroom study with practical experience working for an engineering firm. Although these cooperative, or work-study, programs take longer, they offer significant advantages. Not only does the salary help pay for educational expenses, but the student has the opportunity to apply theoretical knowledge to actual work problems in mechanical engineering. In some cases, the company may offer full-time employment to its co-op workers after graduation.

A graduate degree is a prerequisite for becoming a university professor or researcher. It may also lead to a higher-level job within an engineering department or firm. Some companies encourage their employees to pursue graduate education by offering tuition-reimbursement programs. Because technology is rapidly developing, mechanical engineers need to continue their education, formally or informally, throughout their careers. Conferences, seminars, and professional journals serve to educate engineers about developments in the field.

Certification or Licensing

Engineers whose work may affect the life, health, or safety of the public must be registered according to regulations in all 50 states and the District of Columbia. Applicants for registration must have received a degree from an ABET-accredited engineering program and have four years of experience. They must also pass a written examination.

Many mechanical engineers also become certified. Certification is a status conferred by a professional or technical society for the purpose of recognizing and documenting an individual's abilities in a specific engineering field.

Other Requirements

Personal qualities essential for mechanical engineers include the ability to think analytically, to solve problems, and to work with abstract ideas. Attention to detail is also important, as are good oral and written communications skills and the ability to work well in groups. Computer literacy is essential.

Exploring

One of the best ways to learn about the field is to talk with a mechanical engineer. It might also be helpful to tour an industrial plant or visit a local museum specializing in science and industry. Public libraries usually have books on mechanical engineering that might be enlightening. You might tackle a design or building project to test your aptitude for the field. Finally, some high schools offer engineering clubs or organizations. Membership in the Junior Engineering Technical Society (JETS), a national organization, is suggested for prospective mechanical engineers.

Employers

Most mechanical engineers work in manufacturing, employed by a wide variety of industries. For example, manufacturers of industrial and office machinery, farm equipment, automobiles, petroleum, pharmaceuticals, fabricated metal products, pulp and paper, electronics, utilities, computers, soap and cosmetics, and heating, ventilating, and air-conditioning systems all employ mechanical engineers. Others are self-employed or work for consulting firms, government agencies, or colleges and universities.

Starting Out

Many mechanical engineers find their first job through their college or university placement office. Many companies send recruiters to college campuses to interview and sign up engineering graduates. Other students might find a position in the company where they had a summer or part-time job. Newspapers and professional journals often list job openings for engineers. Job seekers who wish to work for the federal government should contact the nearest branch of the Office of Personnel Management.

Advancement

As engineers gain experience, they can advance to jobs with a wider scope of responsibility and higher pay. Some of these higher level jobs include technical service and development officers, team leaders, research directors, and managers. Some mechanical engineers use their technical knowledge in sales and marketing positions, while others form their own engineering business or consulting firm.

Many engineers advance by furthering their education. A master's degree in business administration, in addition to an engineering degree, is sometimes helpful in obtaining an administrative position. A master's or doctoral degree in an engineering specialty may also lead to executive work. In addition, those with graduate degrees often have the option of research or teaching positions.

Earnings

Starting salaries for mechanical engineers average around $34,000 a year, according to the U.S. Bureau of Labor Statistics. Typically, compensation is considerably higher for those with a graduate degree or more experience. Mid-level engineers earn about $52,000. Those with a great deal of experience can earn annual salaries of $85,000 or more.

Like most professionals, mechanical engineers who work for a company usually receive a generous benefits package, including vacation days, sick leave, health and life insurance, and a savings and pension program. Self-employed mechanical engineers must provide their own benefits.

Work Environment

The working conditions of mechanical engineers vary. Most work indoors in offices, research laboratories, or production departments of factories and shops. Depending on the job, however, a significant amount of work time may be spent on a noisy factory floor, at a construction site, or at another field operation. Mechanical engineers have traditionally designed systems on drafting boards, but, since the introduction of sophisticated software programs, design is increasingly done on computers.

Engineering is, for the most part, a cooperative effort. While the specific duties of an engineer may require independent work, each project is typically the job of an engineering team. Such a team might include other engineers, engineering technicians, and engineering technologists.

Mechanical engineers generally have a 40-hour workweek; however, their working hours are often dictated by project deadlines. They may work long hours to meet a deadline, or show up on a second or third shift to check production at a factory or a construction project.

Mechanical engineering can be a very satisfying occupation. Engineers often get the pleasure of seeing their designs or modifications put into actual, tangible form. Conversely, it can be frustrating when a project is stalled, full of errors, or even abandoned completely.

Outlook

More than three out of five mechanical engineers work in manufacturing fields, such as machinery, transportation (including automotive) equipment, electrical equipment, and fabricated metal products. The remainder are largely found in government agencies and consulting firms.

The employment of mechanical engineers is expected to grow about as fast as average through the year 2006, according the Bureau of Labor Statistics. Although overall employment in manufacturing is expected to decline, engineers will be needed to meet the demand for more efficient industrial machinery and machine tools. Employment for mechanical engineers in business and engineering services firms is expected to grow faster than the average, as other industries increasingly contract out to these firms to solve engineering problems. It should also be noted that reductions in defense spending may adversely affect the employment outlook for engineers within the federal government.

For More Information

For a list of engineering programs at colleges and universities, contact:

Accreditation Board for Engineering and Technology, Inc.
111 Market Place, Suite 1050
Baltimore, MD 21202
Tel: 410-347-7700
Web: http://www.abet.ba.md.us

For information on mechanical engineering and mechanical engineering technology, contact:

American Society of Mechanical Engineers
Three Park Avenue
New York, NY 10016
Tel: 212-591-7000 or 800-THE-ASME
Web: http://www.asme.org

For information about careers and high school engineering competitions, contact:

Junior Engineering Technical Society, Inc.
1420 King Street, Suite 405
Alexandria, VA 22314
Tel: 703-548-5387
Web: http://www.asee.org

Numerical Control Tool Programmers

Computer science Mathematics	School Subjects
Mechanical/manipulative Technical/scientific	Personal Skills
Primarily one location Primarily indoors	Work Environment
Apprenticeship	Minimum Education Level
$27,000 to $41,000 to $50,000+	Salary Range
None available	Certification or Licensing
Decline	Outlook

Overview

Numerical control tool programmers, also called *computer numerical control (CNC) tool programmers,* develop numerical control programs to control machining or processing of parts by automatic machine tools, equipment, or systems.

History

One of the earliest attempts to automate machinery occurred in the early 1700s, when a system of punched cards was used to control knitting machines in England. Holes in punched cards controlled mechanical linkages, which directed yarn colors and allowed various patterns to be woven into a piece of material. Automated machinery did not progress much further, though, until the computer was developed in the late 1940s.

The first use of numerical control (NC) of machine tools in manufacturing was in 1947. John Parsons, owner of a helicopter rotor blade manufacturing company, experimented with regulating milling machinery through numerical control. He discovered that parts made through automated control were more accurate than those made through manual control. The U.S. Air Force, which had a need for uniquely shaped machined parts, contracted with Parsons and the Massachusetts Institute of Technology to develop a machine tool that could be programmed to make contoured parts automatically. In 1952, they built the first numerically controlled machine tool. Shortly afterward, Giddings and Lewis, a large machine tool builder, built an NC profiling mill.

By 1958, other companies followed with NCs of their own. Early numerical control machine tools used paper tapes to program machines. Machine commands were standardized and assigned numerical codes. These codes were then sequenced in the order in which the machine was to perform various operations. After these codes were punched onto a paper tape, a machine operator loaded the tape into a tape reader, loaded raw material, and then started the machine. The machine ran automatically. As numerical control technology evolved, plastic tapes replaced paper, and magnetic spots rather than holes were used to represent codes. Unfortunately, this form of numerical control did not handle changes well—a whole new tape had to be created when process modifications were required. This was a slow and tedious process.

By the 1980s, computer numerical control (CNC) began to replace numerical control. CNC programmers now write programs on computers to sequence the various steps a machine needs to complete. Nowadays many machines have computers or microprocessors built into them. Programmers can easily revise the sequence of operations or other elements. In addition, these programs can store information about the machine tool operation, such as number of parts made and dimensions, request additional raw materials, and record maintenance requirements.

Another advancement in numerical control is direct numerical control, in which several machines are controlled by a central computer. This eliminates the need for individual machine control units and gives programmers more flexibility for modification and control.

The use of CNC machine tools has grown steadily during the last decade and is expected to increase in the future. New applications, such as versatile machining centers, are being developed that allow machines to provide multiple capabilities. Engineers and researchers continue to explore ways to improve the speed, precision, and versatility of machine processes through the use of numerical control and other automated processes.

The Job

Numerical control of machine tools is a form of automated fabrication. Tool programmers write the programs that direct machine tools to perform functions automatically. Programmers must understand how the various machine tools operate and know the working properties of the metals and plastics that are used in the process.

Writing a program for a numerically controlled tool involves several steps. Before tool programmers can begin writing a program, they must analyze the blueprints of whatever function is to be performed or item is to be made. Programmers then determine the steps that must be taken and what tools will be needed. After all necessary computations have been made, the programmers write the program.

Programmers almost always use computers to write the programs, and it is more and more common today to use computer-aided design (CAD) systems. Although tool programmers use computers and CAD systems as aids, they do not write the base software themselves.

To ensure that a program has been properly designed, tool programmers often perform a test or trial run. Trial runs help ensure that a machine is functioning properly and that the final product is correct. Some computers run a simulation program to check a specific program.

Requirements

High School

High school courses in computer science, algebra, geometry, trigonometry, and physics are basics to becoming a CNC programmer. Blueprint reading, computer-aided design and drafting as well as English can add to the background you will need. Shop classes in metalworking can provide an understanding of machinery operations.

Postsecondary Training

Many employers prefer to hire experienced machine workers and then give them training either through programming courses offered by a machine manufacturer or a technical school.

Technical schools and community colleges offer courses in CNC tool programming through various programs. Associate's degrees are available in different areas, such as manufacturing technology and automated manufacturing systems. Typical classes include machine shop, numerical control fundamentals, technical mechanics, advanced NC programming, introduction to robotic technology, and computer-assisted manufacturing. Certificate programs, which generally take one to two years to complete, are also available in such areas as drafting/design and manufacturing technology. These programs usually offer a few classes related to numerical control tool programming.

Some workers learn this trade through apprenticeship programs that combine on-the-job training with classroom instruction. Apprenticeship programs last four years and include training in machine operations program writing, computer-aided design and manufacturing, analyzing drawings and design data, and machine operations. Classes include blueprint reading and drawing, machine tools, industrial mathematics, trigonometry, computers, and operation and maintenance of CNC machines.

Other Requirements

Numerical control tool programmers have an understanding of machine tool operations, possess analytical skills, and show a strong aptitude for mathematics and computers. Generally they possess good communication skills so that they can explain to machine operators how to use and adjust their programs and work with manufacturing engineers to understand what functions must be programmed. They are willing to learn new skills and are flexible to changing needs.

Exploring

If you are interested in a career as a tool programmer, you can test your interest and aptitude by taking shop and vocational classes. You might also visit firms that employ numerical control tool programmers and talk directly with programmers. This is an excellent way to gain practical information on what this type of work is like. Summer jobs or part-time work at manufacturing firms and machine shops may provide insight into the job responsibilities.

Employers

Most tool programmers work in cities where factories and machine shops are concentrated. They work for many types and sizes of companies and businesses. Among the largest employers are industries that produce fabricated metal products, industrial machinery and equipment, transportation equipment, and primary metals.

Starting Out

Tool programming generally is not considered an entry-level job; most employers prefer to hire those with experience and technical training. An individual who has completed courses in tool programming, however, may be hired in spite of a lack of experience. Most often though, firms promote skilled machine workers to programming jobs and then pay for their technical training. Other workers enter this field through apprenticeship programs.

Students who are enrolled in a community college or technical school may learn of job openings through their school's job placement services. Prospective programmers also may learn of openings through state and private employment offices and newspaper ads. Students also can apply directly to a manufacturing firm or machine shop that uses tool programmers.

Advancement

Advancement opportunities are somewhat limited for tool programmers. Employees may advance to higher-paying jobs by transferring to larger or more established manufacturing firms or shops. Experienced tool programmers who demonstrate good interpersonal skills and managerial ability may be promoted to supervisory positions.

Earnings

CNC tool programmers earned an average of $41,000 annually in 1997, according to the U.S. Bureau of Labor Statistics. Apprentices begin at about half the journeyworker rate and earn incremental raises as they progress through training. Tool programmers generally work a 40-hour week, although overtime is common during peak periods. Overtime may include evening and weekend work during production periods when the tool programs are operating.

Benefits vary but may include paid vacations and holidays, personal leaves; medical, dental, vision, and life insurance; retirement plans; profit sharing; and tuition assistance programs.

Work Environment

In general, tool programmers work in comfortable surroundings, and their work is less physically demanding than that of those who operate machine tools. Their work areas usually are separated from the noisier production areas. Numerical control tool programmers are highly skilled workers who must be able to work both individually and as part of a team.

Outlook

There were around 8,500 tool programmers employed in the late 1990s. Initially, employment of CNC programmers was made possible by the introduction of new automation, but recent technological advancements are reducing the demand for such workers. Newer, even more advanced technology now allows some programming and minor adjustments to be made on the shop floor by machinists and machine operators rather than CNC programmers. In the future, fewer programmers will be needed to translate part and product designs into CNC machine tool instructions because newer software does this automatically. Employment is also influenced by economic cycles. As the demand for machined goods falls, programmers involved in this production may be laid off or forced to work fewer hours.

For More Information

For information on apprenticeships, contact:

International Union, United Automobile, Aerospace and Agricultural Implement Workers of America
8000 East Jefferson Avenue
Detroit, MI 48214
Attn: Skilled Trades Department
Tel: 313-926-5000
Web: http://www.uaw.org

For information on careers and educational programs, contact the following sources:

National Tooling & Machining Association
9300 Livingston Road
Fort Washington, MD 20744
Tel: 301-248-6200
Web: http://www.ntma.org

Precision Machined Products Association
6700 West Snowville Road
Brecksville, OH 44141-3292
Tel: 216-526-0300
Web: http://www.pmpa.org

Packaging Engineers

Overview

The *packaging engineer* designs, develops, and specifies containers for all types of goods, such as food, clothing, medicine, housewares, toys, electronics, appliances, and computers. In creating these containers, some of the packaging engineer's activities include product and cost analysis, management of packaging personnel, development and operation of packaging filling lines, and negotiations with customers or sales representatives.

Packaging engineers may also select, design, and develop the machinery used for packaging operations. They may either modify existing machinery or design new machinery to be used for packaging operations.

History

Certain packages, particularly glass containers, have been used for over 3,000 years; the metal can was developed to provide food for Napoleon's army. However, the growth of the packaging industry developed during the

Industrial Revolution, when shipping and storage containers were needed for the increased numbers of goods produced. As the shipping distance from producer to consumer grew, more care had to be given to packaging so goods would not be damaged in transit. Also, storage and safety factors became important with the longer shelf life required for goods produced.

Modern packaging methods have developed since the 1920s with the introduction of cellophane wrappings. Since World War II, early packaging materials such as cloth and wood have been largely replaced by less expensive and more durable materials such as steel, aluminum, and plastics such as polystyrene. Modern production methods have also allowed for the low-cost, mass production of traditional materials such as glass and paperboard. Both government agencies and manufacturers and designers are constantly trying to improve packaging so that it is more convenient, safe, and informative.

Today, packaging engineers must also consider environmental factors when designing packaging because the disposal of used packages has presented a serious problem for many communities. The United States uses more than 500 billion packages yearly; 50 percent of these are used for food and beverages, and another 40 percent for other consumer goods. To help solve this problem, the packaging engineer attempts to come up with solutions such as the use of recyclable, biodegradable, or less bulky packaging.

The Job

Packaging engineers plan, design, develop, and produce containers for all types of products. When developing a package, they must first determine the purpose of the packaging and the needs of the end-user and their clients. Packaging for a product may be needed for a variety of reasons: for shipping, storage, display, or protection. A package for display must be attractive as well as durable and easy to store; labeling and perishability are important considerations, especially for food, medicine, and cosmetics. If the packaging purpose is for storage and shipping, then ease of handling and durability have to be considered. Safety factors may be involved if the materials to be packaged are hazardous, such as toxic chemicals or explosives. Finally, the costs of producing and implementing the packaging have to be considered, as well as the packaging material's impact on the environment.

After determining the purpose of the packaging, the engineers study the physical properties and handling requirements of the product in order to develop the best kind of packaging. They study drawings and descriptions of the product or the actual product itself to learn about its size, shape, weight, and color, the materials used, and the way it functions. They decide what kind of packaging material to use and with the help of designers and production workers, they make sketches, draw up plans, and make samples of the package. These samples, along with lists of materials and cost estimates, are submitted to management or directly to the customer. Computer design programs and other related software may be used in the packaging design and manufacturing process.

When finalizing plans for packaging a product, packaging engineers contribute additional expertise in other areas. They are concerned with efficient use of raw materials and production facilities as well as conservation of energy and reduction of costs. For instance, they may use materials that can be recycled, or they may try to cut down on weight and size. They must keep up with the latest developments in packaging methods and materials and often recommend innovative ways to package products. Once all the details for packaging are worked out, packaging engineers may be involved in supervising the filling and packing operations, operating production lines, or drawing up contracts with customers or sales representatives. They should be knowledgeable about production and manufacturing processes, as well as sales and customer service.

After a packaging sample is approved, packaging engineers may supervise the testing of the package. This may involve simulation of all the various conditions a packaged good may be subjected to, such as temperature, handling, and shipping.

This can be a complex operation involving several steps. For instance, perishable items such as food and beverages have to be packaged to avoid spoilage. Electronic components have to be packaged to prevent damage to parts. Whether the items to be packaged are food, chemicals, medicine, electronics, or factory parts, considerable knowledge of the properties of these products is often necessary to make suitable packaging.

Design and marketing factors also need to be considered when creating the actual package that will be seen by the consumer. Packaging engineers work with graphic designers and packaging designers to design effective packaging that will appeal to consumers. For this task, knowledge of marketing, design, and advertising are essential. Packaging designers consider color, shape, and convenience as well as labeling and other informative features when designing packages for display. Very often, the consumer is able to evaluate a product only from its package.

The many different kinds of packages require different kinds of machinery and skills. For example, the beverage industry produces billions of cans, bottles, and cardboard containers. Often packaging engineers are involved in selecting and designing packaging machinery along with other engineers and production personnel. Packaging can be manufactured either at the same facility where the goods are produced or at facilities that specialize in producing packaging materials.

The packaging engineer must also consider safety, health, and legal factors when designing and producing packaging. Various guidelines apply to the packaging process of certain products and the packaging engineer must be aware of these regulations. Labeling and packaging of products are regulated by various federal agencies such as the Federal Trade Commission and the Food and Drug Administration. For example, the Consumer Product Safety Commission requires that safe packaging materials be used for food and cosmetics.

Requirements

High School

During high school, students planning to enter a college engineering or packaging program should take college algebra, trigonometry, physics, chemistry, computer science, mechanical drawing, economics, and accounting classes. Speech, writing, art, computer-aided design, and graphic arts classes are also recommended.

Postsecondary Training

Several colleges and universities offer a major in packaging engineering. These programs may be offered through an engineering school or a school of packaging within a university. Both bachelor of science and master of science degrees are available. It generally takes four or five years to earn a bachelor's degree and two additional years to earn a master's degree. A master's degree is not required to be a packaging engineer, although many professionals pursue advanced degrees, particularly if they plan to specialize in a specific area or do research. Many students take their first job in packaging once they have earned a bachelor's degree, while other students earn a master's degree immediately upon completing their undergraduate studies.

Students interested in this field often structure their own programs. In college, if no major is offered in packaging engineering, students can choose a related discipline, such as mechanical, industrial, electrical, chemical, materials, or systems engineering. It is useful to take courses in graphic design, computer science, marketing, and management.

Students enrolled in a packaging engineering program will usually take the following courses during their first two years: algebra, trigonometry, calculus, chemistry, physics, accounting, economics, finance, and communications. During the remaining years, classes focus on core packaging subjects such as packaging materials, package development, packaging line machinery, and product protection and distribution. Elective classes include topics concentrating on packaging and the environment, packaging laws and regulation, and technical classes on specific materials. Graduate studies, or those classes necessary to earn a master's degree, include advanced classes in design, analysis, and materials and packaging processes.

Certification or Licensing

Special licensing is required for engineers whose work affects the safety of the public. Much of the work of packaging engineers, however, does not require a license even though their work affects such factors as food and drug spoilage, protection from hazardous materials, or protection from damage. Licensing laws vary from state to state, but, in general, states have similar requirements. They require that an engineer must be a graduate of an approved engineering school, have four years of engineering experience, and pass the state licensing examination. A state board of engineering examiners administers the licensing and registration of engineers.

Professional societies offer certification to engineers instead of licensing. Although certification is not required, many professional engineers obtain it to show that they have mastered specified requirements and have reached a certain level of expertise.

For those interested in working with the specialized field of military packaging technology, the U.S. Army offers courses in this field. Generally, a person earns a bachelor of science degree in packaging engineering before taking these specialized courses. The National Institute of Packaging, Handling, and Logistic Engineers has information about the field of military packaging.

Other Requirements

Packaging engineers should have the ability to solve problems and think analytically and creatively. They must work well with people, both as a leader and as a team player. They should also be able to write and speak well in order to deal effectively with other workers and customers, and in order to document procedures and policies. In addition, a packaging engineer should have the ability to manage projects and people.

Exploring

To get firsthand experience in the packaging industry, students can call local manufacturers to see how they handle and package their products. Often, factories will allow visitors to tour their manufacturing and packaging facilities.

Another way to learn about packaging is by observing the packaging that we encounter every day, such as containers for food, beverages, cosmetics, and household goods. Visit stores to see how products are packaged, stored, or displayed. Notice the shape and labeling on the container, its ease of use, durability for storage, convenience of opening and closing, disposability, and attractiveness.

Students may also explore their aptitude and interest in a packaging career through graphic design courses, art classes that include construction activities, and computer-aided design classes. Participating in hobbies that include designing and constructing objects from different types of materials can also be beneficial. Students may also learn about the industry by reading trade publications, such as *Packaging* and *Packaging Digest.*

Employers

Packaging engineers are employed by almost every manufacturing industry. Pharmaceutical, beverage, cosmetics, and food industries are major employers of packaging engineers. Some packaging engineers are hired to design and develop packaging while others oversee the actual production of the packages. Some companies have their own packaging facilities while other companies subcontract the packaging to specialized packing firms. Manufacturing and packaging companies can be large, multinational enterprises that manufacture, package, and distribute numerous products or they can be small operations that are limited to the production of one or two specific products. Specialized packaging companies hire employees for all aspects of the packaging design and production process. Worldwide manufacturing offers career opportunities around the world. The federal government and the armed services also have employment opportunities for packaging engineers.

Starting Out

College graduates with a degree in packaging or a related field of engineering should find it easy to get jobs as the packaging industry continues its rapid growth. Many companies send recruiters to college campuses to meet with graduating students and interview them for positions with their companies. Students can also learn about employment possibilities through their schools' placement services, job fairs, classified advertisements in newspapers and trade publications, and referrals from teachers. Students who have participated in an internship or work-study program through a college may learn about employment opportunities through contacts with industry professionals.

Students can also research companies they are interested in working for and apply directly to the person in charge of packaging or the personnel office.

Advancement

Beginning packaging engineers generally do routine work under the supervision of experienced engineers and may also receive some formal training through their company. As they become more experienced, they are given more difficult tasks and more independence in solving problems, developing designs, or making decisions.

Some companies provide structured programs in which packaging engineers advance through a sequence of positions to more advanced packaging engineering positions. For example, an entry-level engineer might start out by producing engineering layouts to assist product designers, advance to a product designer, and ultimately move into a management position.

Packaging engineers may advance from being a member of a team to a project supervisor or department manager. Qualified packaging engineers may advance through their department to become a manager or vice president of their company. To advance to a management position, the packaging engineer must demonstrate good technical and production skills and managerial ability. After years of experience, a packaging engineer might wish to become self-employed as a packaging consultant.

To improve chances for advancement, the packaging engineer may wish to get a master's degree in another branch of engineering or in business administration. Many executives in government and industry began their careers as engineers. Some engineers become patent attorneys by combining a law degree with their technical and scientific knowledge.

Many companies encourage continuing education throughout one's career and provide training opportunities in the form of in-house seminars and outside workshops. Taking advantage of any training offered helps one to develop new skills and learn technical information that can increase chances for advancement. Many companies also encourage their employees to participate in professional association activities. Membership and involvement in professional associations are valuable ways to stay current on new trends within the industry, to familiarize oneself with what other companies are doing, and to make contacts with other professionals in the industry. Many times, professionals learn about opportunities for advancement in new areas or at different companies through the contacts they have made at association events.

Earnings

Currently, the average starting salary for a packaging engineer with a bachelor's degree is about $35,000 per year. The mid-range salary is $45,000, with packaging engineers easily earning $60,000 or more as they gain experience and advance within a company.

Benefits vary from company to company but can include any of the following: medical, dental, and life insurance; paid vacations, holidays, and sick days; profit sharing; 401(k) plans; bonus and retirement plans; and educational assistance programs. Some employers pay fees and expenses for participation in professional associations.

Work Environment

The working conditions for packaging engineers vary with the employer and with the tasks of the engineer. Those who work for companies that make packaging materials or who direct packaging operations might work around noisy machinery. Generally, they have offices near the packaging operations where they consult with others in their department, such as packaging machinery technicians and other engineers.

Packaging engineers also work with nontechnical staff such as designers, artists, and marketing and financial people. Packaging engineers must be alert to keeping up with new trends in marketing and technological developments.

Most packaging engineers have a five-day, 40-hour workweek, although overtime is not unusual. In some companies, particularly during research and design stages, product development, and the start-up of new methods or equipment, packaging engineers may work 10-hour days or longer and work on weekends.

Some travel may be involved, especially if the packaging engineer is also involved in sales. Also, travel between plants may be necessary to coordinate packaging operations. At various stages of developing packaging, the packaging engineer will probably be engaged in hands-on activities. These activities may involve handling objects, working with machinery, carrying light loads, and using a variety of tools, machines, and instruments.

The work of packaging engineers also involves other, social concerns such as consumer protection, environmental pollution, and conservation of natural resources. Packaging engineers are constantly searching for safer, tamper-proof packaging, especially because harmful substances have been found in some food, cosmetics, and drugs. They also experiment with new packaging materials and utilize techniques to conserve resources and reduce the disposal problem. Many environmentalists are concerned with managing the waste from discarded packages. Efforts are being made to stop littering; to recycle bottles, cans, and other containers; and to use more biodegradable substances in packaging materials. The qualified packaging engineer, then, will have a broad awareness of social issues.

Outlook

The packaging industry, which employs more than a million people, offers almost unlimited opportunities for packaging engineers. Packaging engineers work in almost any industry because virtually all manufactured products need one or more kinds of packaging. Some of the industries with the fastest growing packaging needs are food, drugs, and cosmetics.

The demand for packaging engineers is expected to increase faster than the average for all occupations as newer, faster ways of packaging are continually being sought to meet the needs of economic growth, world trade expansion, and the environment. Increased efforts are also being made to develop packaging that is easy to open for the growing aging population and those persons with disabilities.

For More Information

The following sources can provide information on educational programs and the packaging industry.

Institute of Electrical and Electronics Engineers (IEEE)
1828 L Street NW, Suite 1202
Washington, DC 20036-5104
Tel: 202-785-0017
Web: http://www.ieee.org/usab

Institute of Packaging Professionals
481 Carlisle Drive
Herndon, VA 22070-4823
Tel: 703-318-8970
Web: http://www.iopp.org/

National Institute of Packaging, Handling, and Logistic Engineers
6902 Lyle Street
Lanham, MD 20706
Tel: 301-459-9105

Packaging Education Forum
481 Carlisle Drive
Herndon, VA 22070-4823
Tel: 703-318-8970
Web: http://www.pakinfo-world.org

Packaging Machinery Manufacturers Institute
4350 North Fairfax Drive, Suite 600
Arlington, VA 22203
Tel: 703-243-8555
Web: http://www.packexpo.com

This Web site lists U.S. and international associations related to the packaging industry.

NAPCO Packaging Association Links from North American Publishing Co.
Web: http://www2.packageprinting.com/packaginglinks.html

Petroleum Refining Workers

Overview

Crude oil and gas are of little use. The value of the crude lies in the refined oils, fuels, and thousands of other products that can be created from it. *Petroleum refining workers* design, test, operate, and maintain equipment and processes that purify crude petroleum into useful, marketable products.

History

Petroleum comes from ancient rock under the ground, where the organic matter of plants and animals has accumulated over millions of years and, through pressure and heat, changed into oil. In early times, petroleum was obtainable only after it had seeped through the earth's surface into above-ground pools. The first oil well was drilled in 1859 because people were looking for a replacement for whale oil, which was difficult to get, to light their lamps. For this they needed kerosene, which they got by boiling the crude in a kind of still. Other products of this early refining were regarded as waste and thrown away including the gasoline we now depend on to fuel our cars.

By the early 1900s the advent of the internal combustion engine to propel "horseless carriages" provided a use for gasoline, as automakers adapted engines to use this practical fuel. And petroleum eventually replaced coal as the major energy source for heating.

Besides motor and heating fuels, petrochemicals that are used in making dyes, cosmetics, plastics, synthetic rubber, and thousands of other products are produced through refining, thanks to a conversion process developed in 1913 by chemists William M. Burton and Robert E. Humphreys.

The rise of the industry was greatly accelerated by World War II. The suddenly increased demand for high-octane aviation gasoline led to a surge in refinery capacity. Also, the need for synthetic rubber required the development of a large-scale technology for producing benzene, butylenes, and other petroleum derivatives. The war also created a demand for many other petrochemical products, including nylon for parachutes and polyethylene to protect electric cables in radar equipment. At the end of the war, pent-up consumer demand kept plants running and petrochemical sales growing at more than 10 percent for the next 20-odd years.

By the late 1960s, Europe and Japan's industrial growth had narrowed the U.S. lead in petrochemicals. In the 1970s, oil-rich nations such as Saudi Arabia, Canada, and Iran began building their own petrochemical plants. By the early 1980s, world trade in petrochemicals had greatly expanded, and today most countries have some petrochemical production facilities.

Though updated, some basic processes and jobs have been performed since petroleum was first refined, while other jobs have been created as a result of the development of new processes; for example, the position of catalytic cracking unit operator was born with the development of this conversion process in the 1930s. Many processes and people are involved in the refining of crude oil, with changes and advances being made as researchers develop new methods.

The Job

The refining world is a highly technical one that, to the untrained eye, appears to be a strange maze of towers, pipes, and tanks. Actually, it is an organized and coordinated arrangement of manufacturing processes designed to produce physical and chemical changes in crude oil and gas, resulting in saleable products.

The types of crude to be processed and the requirements of the market determine the techniques used in a refinery. More than a hundred different types of crudes are internationally traded, and a modern refinery may process as many as twenty grades in the course of a year.

Three basic refining procedures are separation, conversion, and treatment. The first procedure separates the crude oil into different parts, or fractions, and is usually accomplished by distillation in tall tanks called fractionating columns. In this process, the oil is heated in pipes in a huge furnace. The heated oils produce vapors, which pass into the fractionating column, where they condense into materials with different properties. Pipes that are set at different levels in the towers draw off the various fractions, which rise to different heights depending on their weight and other characteristics. Distillation is continuous, with hot crude oil flowing in near the base of the column and the separate fractions flowing out at each level. The lighter fractions, like liquefied petroleum gases (LPG), gasoline, and naphtha, a major feedstock for the chemical industry, are tapped off from the top of the tower. Heavier fractions, like kerosene (jet fuel), diesel fuel, and heating oil, are drawn off around the midsection of the towers; very heavy materials, such as fuel oils or residues, remain in the bottom sections.

Products from the fractionating columns are straight-run products. They may be ready for the market, or they may be treated to remove impurities. Often they go through conversion techniques to chemically change their makeup. For example, fuel oil, one of the heaviest fractions, may account for between one-third and one-half of the yield from distillation, whereas demand from customers is predominantly for the lighter fractions, such as gasoline. This is especially so in the United States, with all its cars. Heavy oil fractions can be converted to high-octane gasoline through a heat and pressure process called cracking to help meet market demand.

Distillation and cracking also produce petrochemicals. At petrochemical plants, which may be separate facilities or, increasingly, part of the refinery complex, these first-stage petrochemicals are changed via conventional chemistry into secondary petrochemicals, which in turn are transformed into the medicines, fabrics, cosmetics, detergents, and other everyday products used at home and work.

In recent years, many refineries have invested considerably in conversion facilities, installed computers to process refinery operations, and introduced energy management plans. At the beginning of 1996, there were about 165 operating refineries in the United States, employing 101,600 workers in a wide range of jobs. In general, those jobs can be classified into four broad categories: operations, maintenance, engineering and scientific support, and supervision. Following are a number of examples of jobs in each category.

Operations workers run the variety of machines that refine petroleum. Control panel operators work the gauges that regulate the temperature, pressure, rate of flow, and tank level in petroleum refining and petrochemical processing units. They observe and regulate meters and instruments to process petroleum under specified conditions. When the fractionating process is complete, *treaters* are responsible for controlling the equipment that removes impurities and improves the quality of gasoline, kerosene, and lubricants; to perform their jobs, treaters use steam, clay, hydrogen, solvents, and chemicals. *Clay roasters* use a kiln to clean and treat clay that has been used to treat oil.

Certain refined oils are blended to achieve specific qualities or make specific fuels. Various types of workers are needed for such operations. *Compounders* add antioxidants, corrosion inhibitors, detergents, and other additives to enhance lubricating oils. *Blenders* mix gasoline with chemicals, lead, or distilled crude oil to make specified commercial fuel. *Grease makers* heat oils with fat, soda, water, dye, and mineral oils to produce various grades of lubricating grease.

Involved in the operation of petrochemical processing plants are *paraffin plant operators,* who work with filter presses to separate paraffin oil distillate from paraffin wax. Paraffin plant *sweater operators* operate tanks that heat and cool substances to separate liquid from processed paraffin distillate. *Lead recoverers* at naphtha-treating plants operate centrifuge machines that separate lead compounds from a naphtha solution used to treat gasoline.

Other operations workers include *oil-recovery-unit operators,* who separate recoverable oil from refinery sewage systems, and *refinery laborers,* who prepare work sites, load and unload equipment, dump ingredients for mixing into machines, and perform many other tasks. In addition, the industry employs many workers to load and drive delivery trucks.

Maintenance workers make up more than half of all refinery employees. They keep the workplaces safe and the machinery in working condition. Fire marshals are needed because of the flammable nature of petroleum and its by-products. *Refinery marshals* coordinate firefighters' activities, inspect equipment and workplaces to make sure they meet fire regulations, order fire drills, and direct fire fighting and rescues in the event of a refinery fire. *Mechanical inspectors* inspect tanks, pipes, pipe fittings, stills, towers, and pumps for defects and report the need for repairs. They use special instruments to measure the thickness of tower walls and pipes and to determine rates of corrosion and decay. *Line walkers* patrol pipelines to look for leaks. *Salvagers* are responsible for fixing defective valves and pipe fittings.

Other maintenance workers include *gas-regulator repairers,* who fix and install equipment that controls the pressure of gases used in petroleum refining; *meter testers,* who ensure the correct functioning of the meters that indicate the flow and pressure of gases, steam, and water; and *electricians,* who repair and maintain refineries' electrical systems, including motors, transformers, wiring, switches, and alarms. Other workers, including *tube cleaners,* keep equipment clean so it will perform at required standards. *Tankcar inspectors* examine the wheels, bearings, brakes, and safety equipment of refining tank cars to prevent catastrophic accidents. *Construction and maintenance inspectors* inspect petroleum-dispensing equipment at distribution plants.

Engineers, chemists, and other scientific support staff are involved in tasks that require more theoretical knowledge than those done by operations and maintenance workers. These workers are often responsible for the actual design and development of refining plants. They also devise ways to treat and improve products, and they develop, improve, and test refinery processes.

There is also a variety of managers and supervisors who work in the petroleum refining industry. *Contract managers* negotiate the purchase and delivery of crude oil, and they negotiate sales of refinery products. *Purchasing managers* procure chemicals, catalysts, piping, valves, motors, pumps, and many other commodities for use in the refinery complex. Services include contracting for construction work, waste disposal, janitorial, catering, photography, and transportation. *Bulk plant managers* manage storage and distribution facilities for petroleum products. *Dispatchers* regulate the flow of products through processing, treating, and shipping departments. In addition, most operations at refineries are headed by *supervisors,* who coordinate workers' activities, plan production schedules, and oversee processes.

Requirements

High School

Because of the diversity of functions, qualifications differ greatly. Petroleum refining workers who are involved in operations and maintenance need at least a high school diploma, and a year or two at college or a technical school is very helpful and highly desirable to most employers. Maintenance workers may also be required to have special training in repairing certain kinds of equipment.

Operations workers learn most of their skills on the job. High school and college-level courses in chemistry, physics, and mathematics, however, should help them understand and learn these skills more readily. As computers are now widely used in refinery operations, those who take computer science classes will discover that computer skills are not only valuable, but essential for some jobs.

Postsecondary Training

Engineers and scientists need college degrees from accredited institutions. Chemical and laboratory technicians may be graduates of a two-year program, but engineers and scientists must have a minimum of a bachelor's degree, and many also have a master's degree and doctorate. High school students preparing for such college study should take math, chemistry, physics, biology, drafting, and computer applications, as well as English classes.

Other Requirements

Most work in refineries requires a high degree of precision and accuracy, and many positions require knowledge of intricate machine operations. Workers should be alert, attentive, and quick-thinking, and able to work under pressure.

Exploring

Talking with someone who has worked in a refinery would be a very helpful and inexpensive way of exploring this field. One good way to find an experienced person to talk to is through online computer services.

Another way to learn about petroleum refining occupations is checking school or public libraries for books on the petroleum refining and petrochemical industries. Industry unions, to which most operations and maintenance refinery workers belong, are also good sources of information about this type of work; one of the largest such unions is the Oil, Chemical and Atomic Workers International Union. Other resources include trade journals, high school guidance counselors, the placement office at technical or community colleges, and the associations listed at the end of this article.

Most refineries are very open to high schoolers taking free plant tours and provide information to students. Contact the public relations department of a nearby refinery to arrange a tour or to request information.

Some summer and other temporary jobs in refineries are available, and they provide a very good way of finding out about this field. Because refineries are in operation 24 hours a day, late-shift work may also be available to those exploring the industry. Temporary workers can learn firsthand the basics of oil refining operations, equipment maintenance, safety, and other aspects of the work.

Employers

Refining workers can find employment at large and small refineries and petrochemical companies.

Starting Out

High school or technical school graduates should fill out applications at employment offices of refineries. State employment offices and local offices of trade unions such as the Oil, Chemical and Atomic Workers International can also be contacted for employment information. Apprenticeships are often available to teach workers specialized skills. These programs may take as many as four to five years to complete, and they combine formal classroom instruction with practical experience.

College and university students may interview on campus with companies that have recruitment programs. In addition, college placement offices usually have information on positions and internships.

All job seekers can increase their chances of success by familiarizing themselves with the concerns and activities of potential employers. Many companies make their annual reports available, which can be found in the business section at many libraries, along with trade and professional magazines and journals.

Advancement

Advancement in the petroleum refining industry depends on experience, type of position, and place of employment (larger refineries tend to have greater advancement opportunities). Seniority plays a large role in promotions for operations and maintenance workers, since they belong to unions. Operations workers may move to a more responsible, higher-paying position such as grease maker, which requires long experience and specialized knowledge. They may also move up to supervisory jobs.

An advancement path for a maintenance worker might be machinist, to machinery assistant supervisor, to maintenance mechanic superintendent with responsibility for planning and budgeting repairs. With additional engineering and business administration schooling, the worker could further advance to senior machinery reliability specialist, responsible for finding the best, most cost-effective way to improve refinery equipment reliability, designing and installing new refinery equipment, and training supervisors on that equipment.

Engineers and scientists may be promoted to positions involving management and supervision, in which they are responsible for supervising many workers and taking charge of entire refining processes. Someone with a bachelor's in chemistry, for example, might start out as a junior research chemist, then move up to senior research chemist, to product, process, and environmental quality chemist, with responsibility for sampling and analyzing products, processes, and environmental effects of refining. From there the chemist could progress to chief plant chemist.

This person could also be an industrial hygiene chemist, performing occupational health, air, and water pollution control sampling and analyses, continuing on to industrial hygienist, with responsibility for industrial hygiene programs including performing safety investigations and recommending and designing controls to maintain healthful working conditions. Ultimately, an industrial hygienist could be supervisor of safety and industrial hygiene, managing an entire department. For this career path, certification in hazardous materials and industrial hygiene would be necessary, as well as additional schooling in safety engineering.

In general, persons with good judgment, who can act effectively in emergencies, and who are willing to learn new skills and obtain additional schooling when necessary, should be in a good position to receive promotions as they become available.

Earnings

Because of the diversity of jobs, salaries differ greatly. Salaries also vary with geographic location, experience, education, and employer. In 1995, refinery maintenance and operations workers in nonsupervisory positions earned an average of $19 per hour, according to the Oil, Chemical and Atomic Workers International Union, although salaries are dependent on the worker's job type and level of experience. These workers also receive premium pay for evening and night shifts as well as overtime pay.

Salaries for engineering and scientific support positions are higher. Entry-level mechanical engineers earn about $35,000 to $40,000 per year; experienced engineers, up to $80,000, according to an industry source. A 1995 salary survey from the American Chemical Society gave a mean salary of $39,799 for chemical engineers with a bachelor of science degree and two to four years of experience, ranging up to $81,255 for 40 or more years of experience. For chemical engineers with a master's, mean salaries ranged from $49,747 to $86,766, depending on experience; with a doctorate, from $59,010 to $110,410, depending on experience. The same salary survey showed mean salaries for petroleum industry chemists of $38,560 to $105,864, depending on experience and education. Entry-level chemical technicians had a mean salary of $24,430, going up to $43,292 for 30 or more years of experience.

Workers in this industry have paid vacations, pensions, and health and life insurance.

Work Environment

Plants operate 24 hours a day, year-round, resulting in an air of urgency that is a permanent part of the job. Overtime is often involved, and workers may be requested to work nights, evenings, weekends, or holidays, depending on seniority. Extreme pressure occurs during shutdown of selected operating units for maintenance overhaul.

Many refinery operations are located in fairly remote areas. Conditions in locations such as Alaska and the North Sea can be quite rugged during winter months. Refinery work often takes workers outside to check equipment. Those who work indoors usually encounter clean, modern conditions, usually some distance away from the equipment or the operations.

Outlook

Although domestic production of oil has declined, the United States continues to import much crude oil. In addition, demand for petroleum products,
while not growing much, is holding steady, and employment of refinery
workers is expected to be steady. A few thousand jobs will open up each year
as workers retire, transfer, or otherwise leave the industry.

However, the refining industry is much leaner than it used to be. Due to
less domestic oil production and environmental regulations, close to half of
the petroleum refineries operating in 1982 have since closed. U.S. refiners
have had to reevaluate their operations closely in light of federal and state
environmental regulations. Large companies with multiple refining operations have had to commit substantial resources to plant additions and recon-
figurations, product reformulations, and research and development of processing technologies to comply with regulations. The required investments
are greater on a per-barrel basis for smaller refineries, many of which didn't
have the resources to meet new environmental standards and were forced to
shut down. Older refineries too were particularly hard hit.

Refining operations and capital investment will continue to be affected
by pending technological standards to reduce emissions. The increases in
environmental costs come at a time of little growth in overall demand for
petroleum products. Therefore, the trend for certain refineries to close down
partially or entirely rather than upgrade facilities to meet the new standards
is likely to continue.

Besides considering the industry overall, those contemplating a career in
petroleum refining need to be aware of what individual companies are doing.
Some companies are hiring workers, while many others are downsizing. The
discrepancy is due to some companies moving away from domestic refining
operations, while others are concentrating on that side of the business. In the
past, it was common for oil companies to be involved in all areas of the
industry—exploration, drilling and production, transportation, refining, and
distribution/marketing—in both the domestic and worldwide arena. Now
companies are trying to narrow their place in the market based on their
strengths instead of trying to do it all.

Closely linked to the oil industry is the petrochemical industry, which is
currently the fastest-growing industrial chemical field and produces the
largest number of new chemicals. Because most of its products are considered
hazardous at some stage in their manufacture, the petrochemical industry has
also been profoundly affected by government environmental regulations.

Automation and computerization are changing the face of the refinery
employment population. Refinery personnel recruiters are more attracted to
applicants with computer, chemistry, engineering, and mechanical back-

grounds than they are to unskilled workers. While automation may be decreasing the number of jobs for operations workers, jobs will open up for maintenance workers such as pipefitters, electricians, machinists, and repair personnel. Employment will remain fairly static among professional, technical, and administrative workers.

For More Information

This trade association represents employees in the petroleum industry. Free videos, fact sheets, and informational booklets are available to educators.

American Petroleum Institute
1220 L Street, NW
Washington, DC 20005
Tel: 202-682-8000
Web: http://www.api.org

For information about careers, salaries, and employment trends, contact:

American Chemical Society
Department of Career Services
1155 16th Street, NW
Washington, DC 20036
Tel: 800-227-5558
Web: http://www.acs.org

Oil, Chemical and Atomic Workers International Union
Public Relations
PO Box 281200
Lakewood, CO 80228-8200
Tel: 303-987-2229

For information about JETS programs, products, and engineering career brochures (all disciplines), contact:

Junior Engineering Technical Society, Inc.
1420 King Street, Suite 405
Alexandria, VA 22314-2715
Tel: 703-548-5387
Email: jets@nae.edu
Web: http://www.jets.org

For Opportunities for Performance, *a booklet describing professional jobs at Phillips Petroleum Company, contact:*

Phillips Petroleum Company, Employment & College Relations
180 Plaza Office Building
Bartlesville, OK 74004
Tel: 918-661-6385
Web: http://www.phillips66.com

For Oil, *a booklet describing all phases of the oil industry, and information on careers with Shell, contact:*

Shell Oil Company
External Affairs
PO Box 2463
Houston, TX 77252-2463
Tel: 713-241-6161
Web: http://www.shell.com

Pharmaceutical Industry Workers

Overview

Pharmaceutical industry workers are involved in many aspects of the development, manufacture, and distribution of pharmaceutical products. *Pharmaceutical operators* work with machines that perform such functions as filling capsules and inspecting the quality and weight of tablets. *Pharmaceutical supervisors and managers* oversee research and development, production, and sales and promotion workers. *Pharmaceutical sales representatives* sell and distribute pharmaceutical products and introduce new items to pharmacists, retail stores, and medical practitioners.

History

The oldest known written records relating to pharmaceutical preparations—5,000 years old—come from the ancient Sumerians. Other ancient cultures, such as the Indians and Chinese, used primitive pharmaceutical applications to eradicate evil spirits, which they believed to cause evil in the body. The Babylonians, Assyrians, Greeks, and Egyptians also compounded early pharmaceuticals in hope that they would rid the body of disease (which they believed was caused primarily by sinful thoughts and deeds).

Professions in pharmacy began to be established in the 17th century, after the first major list of drugs and their applications and preparations was compiled. The discoveries of the anesthetics, morphine (first used in 1806), ether (1842), and cocaine (1860), were among the first pharmaceutical advancements to significantly benefit humankind. Since then, numerous vaccines have cured sickness and disease and in effect have helped people live longer, healthier lives.

In 1852 in the United States, the American Pharmaceutical Association was formed to help those in the pharmaceutical field organize their professional, political, and economic goals (the Pharmaceutical Manufacturers Association replaced the APA in 1958). Government intervention in the pharmaceutical industry began in 1848, and in 1931 the Food and Drug Administration (FDA) was formed to provide legal regulation and monitoring of the pharmaceutical industry. As the industry became increasingly regulated and organized, qualified, trained workers were sought to professionally develop, produce, package, and market pharmaceutical products. These workers, known collectively as pharmaceutical industry workers, possess a variety of skills, responsibilities, and education levels and continue to actively work to improve the quality and length of our lives.

The Job

The pharmaceutical, or simply, the drug industry, has four main divisions: research and development, production, administration, and sales.

Research and development professionals create new drug products and improve existing ones. The products they design are manufactured by production workers called, as a whole, pharmaceutical operators. Many of these employees work on production lines, tending equipment that measures, weighs, mixes, and granulates various chemical ingredients and components, which are then manufactured into such forms as pills and capsules. Often these employees inspect the finished goods, looking for such inconsistencies as broken tablets and unfilled capsules.

There are a number of specific job designations in the realm of production. *Capsule filling machine operators* run machines that fill gelatin capsules with medicine. They scoop empty capsules into a loading hopper and medicine into a filling hopper. After the filled and sealed capsules are ejected by the machinery, these operators inspect the capsules for proper filling and for evidence of breakage. They may also spot-check individual capsules or lots by comparing their weight with standardized figures on a weight specification sheet. This process is used for certain antihistamines, vitamins, and general pain relievers, for example.

Ampule and vial fillers work with glass tubes and plastic and glass containers with rubber stopules that are filled with medicine and then sealed. The process for filling is similar to that for the capsule filler; however, the operator must adjust gas flames to the appropriate temperature so that the tubes are completely sealed. They also count and pack readied ampules and vials for shipment. (Vials and syringes have recently become the primary containers for liquid drug production in the United States.)

Ampule and vial inspectors use magnifying glasses to check for cracks, leaks, and other damage. They keep records of inspected cartons, as well as damaged or flawed products.

Granulator machine operators operate mixing and milling machines that are equipped with fine blades that mix ingredients and then crush or mill them into powdered form so that they can be formed into tablets. They are responsible for weighing and measuring each batch, blending the ingredients with the use of machinery, and adding alcohol, gelatins, or starch pastes to help the pill keep its form. They then spread the mixture on trays which they place into an oven or steam dryer set at a predetermined temperature. At the conclusion of the heating process, they check each batch for dryness levels, size, weight, and texture.

Coaters operate machines that cover pills and tablets with coatings that flavor, color, or preserve the contents.

Fermenter operators oversee fermenting tanks and equipment, which produce antibiotics and other drugs. Operators start the mixing tanks, add ingredients, such as salt, yeast, and sugar, and transfer the mixture to a fermenting tank when it is ready. They are responsible for monitoring the temperature in the tanks, for adding precise amounts of liquid antibiotic, water, and foam-preventive oil, and for measuring the amount of solution so that it may be transferred to another tank for additional processing.

There are also a vast number of laborer professions involved in the production area of the pharmaceutical industry. *Hand packers and packagers* remove filled cartons from conveyor belts and transport other finished pharmaceutical products to and from shipping departments. *Industrial machinery mechanics* ensure that all machinery is working properly and at optimum production capacity.

The third major branch of this industry comprises administrative positions. *Production managers* direct workers in the manufacturing field by scheduling projects and deadlines. These employees also oversee factory operations and enforce safety and health regulations, monitor efficiency, and plan work assignments. They also direct and schedule assignments for the shipping department, which packs and loads the pharmaceutical products for distribution.

The finished products are marketed by the sales branch. *Service and sales representatives* supply pharmaceutical drugs and related products to hospitals, independent medical practitioners, pharmacists, and retail stores. Telephone calls and office visits allow the representatives to keep in contact with buyers, monitor supplies, and introduce new products. Often, reps supplement free samples of new products with printed literature when available. Sales reps may choose to promote, for example, certain vitamins and other nutritional supplements, pain relievers, and general health care supplies. Jim Batastini works as a sales representative for the pharmaceutical industry; much of his work entails calling on physicians within a certain geographic area. "The purpose of my visits with these physicians," he says, "is to provide the latest clinical information relevant to our products and how they can best be used to manage different disease states."

Requirements

High School

To prepare for a job in sales or administration, you should take courses in speech and English, to develop your communication skills. You should also take science courses, including biology and chemistry, so that you'll have some insight into pharmaceutical research and development. If you're interested in a job as a production worker, take courses in math and science. You should also take courses, such as voc-tech, that will give you some background in machine work and engineering.

Postsecondary Education

Most employers offering production jobs require at least a high school diploma or the equivalent. Certain labor positions also require technical or vocational training.

Some pharmaceutical companies offer on-the-job training to nonprofessional workers. Employees in sales may be required to have sufficient training or a background in pharmacology as well as in sales and marketing, whereas certain administrative positions require course work in liberal arts, data processing, and business administration. Various types of pharmaceutical training are also available in the military. Information about pharmaceutical careers in the armed forces can be obtained by contacting your nearest military recruitment office.

Other Requirements

Pharmaceutical industry workers on the whole must be alert, dependable, and possess good communications skills, both oral and written. Workers can expect to interact with all divisions and levels of employees—strong communications skills promote faster and more accurate production. Production workers must be physically fit, mentally alert to oversee production lines and processes, and have the temperament to work at sometimes repetitive tasks. Administrative and managerial workers must be decisive leaders with empathy for workers at all levels of education and responsibility. Sales and marketing workers need good people and persuasive skills in order to effectively promote products. "You should have the ability to learn a large volume of technical material," Batastini says, "and have the ability to assimilate the information and concisely communicate it to medical professionals."

Exploring

If your high school has a vocational training program, look into taking a class that will prepare you for production work; a local community college may also have such a course. You should consider contacting trade organizations such as the American Foundation for Pharmaceutical Education, whose objective is to improve pharmaceutical educational programs and student performance. In addition, science-related clubs and social organizations often schedule meetings and professional lectures and offer career guidance as well.

To prepare for a sales career, you might be able to find part-time work in a pharmacy. Working for a pharmacy, you can learn about the drug manufacturers, the most-prescribed drugs, and other information about the industry. You may also meet sales representatives, and have the opportunity to read the promotional materials distributed by drug companies.

Employers

Production workers and sales representatives work for pharmaceutical companies that manufacture prescription and over-the-counter products. These companies include Johnson & Johnson and Bristol-Myers Squibb. A small percentage of industry workers are employed with companies that make the biological products that are used by manufacturers in the production of drugs.

Starting Out

College-trained applicants often benefit from placement services provided by the student services division of their schools. Applicants can also apply directly to pharmaceutical companies or through school contacts with professional organizations. In addition, newspapers and professional trade publications list job opportunities that are offered in each division and level of the industry.

"If you're interested in pharmaceutical sales," Batastini says, "a strong science background with good academic standing will probably be required in the future. And networking with people in the industry is a great way to get your foot in the door."

Advancement

There are many advancement opportunities for pharmaceutical industry workers. Production workers may advance to managerial positions and learn how to operate more sophisticated machinery. Laboratory assistants and research assistants may prepare for advancement with additional education and be promoted to new research projects and duties. Administrators may become supervisors, executives, sales managers, or marketing executives.

There are always possibilities for advancement for employees who are willing to develop new skills and take on more responsibilities. Many positions, however, require additional, formal training. "The industry maintains a high level of continuing education requirements," Batastini says.

Earnings

Because the pharmaceutical industry is a large field, earnings vary tremendously and depend on the worker's position, educational background, and amount of work experience. However, some generalizations can be made about certain wages.

Production workers average approximately $14 per hour, though the wage range for these employees is broad, depending on the size of the firm, the shift to which the worker is assigned, years at the company, and the geographic location of the plant. Overtime compensation is usually equal to time and a half or double time.

According to *The National Business Employment Weekly,* those in pharmaceutical sales make between $34,000 a year and $55,000, depending on the position. The median is about $43,000 a year.

All full-time workers, regardless of their work specialty, receive paid vacations, medical and dental insurance, paid sick and personal days, pension plans, and life insurance. Some workers may also be offered profit-sharing, savings plans, and reimbursement for job-related education.

Work Environment

Production workers average 45-hour work weeks and eight hours per shift; at some pharmaceutical firms, however, shifts may run round the clock, meaning that some employees work a variety of shifts. Production workers often work in chemical factories, which are well ventilated and offer good lighting but may be noisy and crowded. These workers may have to package products and load them onto trucks or docks by hand or with forklifts. Machinery operators may stand much of their shift. Laborers and packagers frequently walk, stand, bend, and lift in the course of their day. They may be required to operate machinery to lift heavy or bulky material. Ampule and vial fillers wear special clothing, such as complete face and body coverings, to maintain sterile conditions. Safety equipment is required for hazardous tasks of all types.

Administrators work in office environments that are often modern, neat, and have good lighting and ample work spaces. They often bring work home with them or have late meetings with other staff members.

Advertising and sales workers travel considerable distances to hospitals, pharmacies, and physicians' offices. They may go to other cities or even other countries to promote their product line. Batastini says the work sometimes requires 80 hours per week. "But it's an ever evolving field," he says, "so you never get bored with the subject matter."

Outlook

As the U.S. population continues to include increasing numbers of older people, the pharmaceutical industry is expected to grow to accommodate medical needs. In addition, technological developments continue to be pursued in many scientific endeavors, including the creation of new drugs for the treatment of such widespread diseases as AIDS and cancer. The overall employment outlook for workers in the pharmaceutical industry is thus considered very good and is anticipated to continue at a growing pace at least through the first years of the 21st century.

Many pharmaceutical manufacturing companies are investigating growth in health-related areas, such as cosmetics, veterinary products, agricultural chemicals, and medicinals and botanicals. Dietary supplements account for $12 billion a year in sales, and the demand for herbal products grew from about $1.6 billion a year to more than $3 billion a year in 1998.

In production positions there will be a decline for machine operators as more machinery becomes automated. Inspectors, testers, and graders may also see a decline in jobs as a result of downsizing. Expected increases in production positions include industrial machinery mechanics, hand packers and packagers, and granulator machine operators. Sales and marketing personnel will be needed to educate buyers about newly approved over-the-counter medications, generics, biopharmaceuticals, and other new products. Some companies have experimented with selling products directly in magazines, which has increased sales for some drugs. But these increased sales may not be enough to account for the expense of advertising.

For More Information

For more information about the pharmaceutical industry, contact:

American Foundation for Pharmaceutical Education
One Church Street, Suite 202
Rockville, MD 20850
Tel: 301-738-2160

National Association of Pharmaceutical Manufacturers
320 Old Country Road
Garden City, NY 11530-1752
Tel: 516-741-3699
Web: http://www.napmnet.org

For information about student membership and publications, and for news about the industry, visit the APhA Web site, or contact:

American Pharmaceutical Association
2215 Constitution Avenue, NW
Washington, DC 20037-2985
Tel: 202-628-4410
Web: http://www.aphanet.org

Plastics Products Manufacturing Workers

	School Subjects
Mathematics Technical/Shop	
	Personal Skills
Following instructions Mechanical/manipulative	
	Work Environment
Primarily indoors Primarily one location	
	Minimum Education Level
High school diploma	
	Salary Range
$19,000 to $28,000 to $40,000	
	Certification or Licensing
Voluntary	
	Outlook
Faster than the average	

Overview

Plastics products manufacturing workers mold, cast, and assemble products made of plastics materials. The objects they make are almost without number. They include dishes, signs, toys, insulation, appliance parts, automobile parts, combs, gears, bearings, and many others.

History

Thermoplastics, plastics that soften with heat and harden when cooled, were discovered in France in 1828. In the United States in 1869, a printer named John Wesley Hyatt attempted to create an alternative material to supplement ivory in billiard balls. He experimented with a mixture of cellulose nitrate and camphor, creating what he called celluloid. His invention, patented in

1872, brought about a revolution in production and manufacturing. By 1892, over 2,500 articles were being produced from celluloid. Among these inventions were piano keys, false teeth, and the first movie film. Celluloid did have its drawbacks. It could not be molded and it was highly flammable.

It was not until 1909, however, that the Belgian-American chemist Leo H. Baekeland produced the first synthetic plastic. This product replaced natural rubber in electrical insulation and was used for phone handsets, and automobile distributor caps and rotors, and is still used today. Other plastics materials have been developed steadily. The greatest variety of materials and applications, however, came during World War II, when the war effort brought about a need for innovation in clothing, consumer goods, transportation, and military equipment.

Today, plastics manufacturing is a major industry whose products play a vital role in many other industries and activities around the world. It is difficult to find an area of our lives where plastics do not play some role. Major users of plastics include the electronics, packaging, aerospace, medical, and housing and building industries. The plastics industry also provides the makings for a large variety of consumer goods. Appliances, toys, dinnerware, luggage, and furniture are just a few products that require plastics.

Plastics products manufacturing workers have always been needed in the production of plastic. Their job responsibilities and skills have changed and grown more specialized as new productions processes and materials have come into widespread use.

The Job

Plastics are usually made by a process called polymerization, in which many molecules of the same kind are combined to make networks of giant particles. All plastics can be formed or shaped; some become pliable under heat, some at elevated room temperatures. When treated, some plastics become hard, some incredibly strong, some soft like putty.

Plastic objects are formed using several different methods. Each method produces a different type of plastic. In compression molding, plastics compounds are compressed and treated inside a mold to form them. In injection molding, liquid plastic is injected into a mold and hardened. Blow molding is like glass blowing—air is forced into plastic to make it expand to the inner surface of a mold. In extrusion, hot plastic is continuously forced through a die to make products like tubing. Laminating involves fusing together resin-soaked sheets, while the calender process forms sheets by forcing hot plastic between rollers. Finally, in fabrication, workers make items out of solid plastic pieces by heating, sawing, and drilling.

Marvin Griggs works for a company called Centro, in Springdale, Arkansas. "We're a rotational molder for plastic products," Griggs says. "We make custom parts for companies like John Deere. We don't produce our own product. They send the mold, we build the parts." Griggs is part of a four-person crew running one of the machines—the *machine operator* and *assistant operator* pour resin into the molds, which is then placed into the oven, and then the cooler. They open the molds and remove and inspect the part for warping, or some other defect. The *trimmer* then trims the line, cutting off the plastic flange. "If the part needs holes cut into it, or fixtures put in it," Griggs says, "they pass it down to me." The tools he uses include pneumatic hand tools, routers, and a large tank. "The parts are dunk-tested to make sure they're sealed, and there are no holes. We have a tight quality control system."

While plastics compounds may be mixed in plastics-materials plants, plastics fabricators sometimes employ blenders, or color mixers, and their helpers to measure, heat, and mix materials to produce or color plastic materials. *Grinding-machine operators* run machines that grind particles of plastics into smaller pieces for processing. *Pilling-machine operators* take plastics powder and compress it into pellets or biscuits for further processing. Other workers are responsible for making the molds (plastic form makers) and patterns (plastics patternmakers) that are used to determine the shape of the finished plastics items. Foam-machine operators spray thermoplastic resins into conveyor belts to form plastic foam.

Many plastics products plants make goods according to clients' specifications. When this is the case, *job setters,* using their knowledge of plastics and their properties, adjust molding machines to clients' instructions. They make such adjustments as changing the die through which the plastic flows, adjusting the speed of the flow, and replacing worn cutting tools when necessary. Then the machine is ready to accept the plastic and produce the object.

Injection molders operate machines that liquefy plastic powders or pellets, inject liquid plastic into a mold, and eject a molded product. Compact discs, toys, typewriter keys, and many other common products are made by injection molding. *Injection workers* set and observe gauges to determine the temperature of the plastic and examine ejected objects for defects.

One common plastic is polystyrene, which when molded using heat and pressure makes cast foam products such as balls, coolers, and packing nests. *Polystyrene-bead molders* operate machines that expand these beads and mold them into sheets of bead board. *Polystyrene-molding-machine tenders* run machines that mold pre-expanded beads into objects. At the end of the molding cycle, they lift the cast objects from the molds and press a button to start the machine again.

Extruder operators and their helpers set up and run machines that extrude thermoplastics to form tubes, rods, and film. They adjust the dies and machine screws through which the hot plastic is drawn, adjust the machine's cooling system, weigh and mix plastics materials, empty them into the machine, set the temperature and speed of the machine, and start it.

Blow-molding-machine operators run machines that mold objects such as bleach bottles and milk bottles by puffing air into plastic to expand it. *Compression-molding-machine operators and tenders* run machines that mold thermosetting plastics into hard plastic objects. Thermosetting plastics are those that harden because of a chemical reaction rather than by heating and cooling.

Casters make similar molded products by hand. *Strippers* remove molded items from molds and clean the molds. Some molded products must be vacuum cured. *Baggers* run machines that perform this task.

Plastic sheeting is formed by *calender operators,* who adjust the temperature, speed, and roller position of machines that draw plastic between rollers to produce sheets of specified thickness. *Stretch-machine operators* stretch plastic sheets to specified dimensions. *Preform laminators* press fiberglass and resin-coated fabrics over plaster, steel, or wooden forms to make plastic parts for boats, cars, and airplanes.

Other common plastics products are fiberglass poles and dowels. *Fiberglass-dowel-drawing-machine operators* mount dies on machines, mix and pour plastics compounds, draw fiberglass through the die, and soak, cool, cure, and cut dowels. *Fiberglass tube molders* make tubing used in fishing rods and golf club shafts.

Plastics that are not molded may be cut into shapes. *Shaping-machine operators* cut spheres, cones, blocks, and other shapes from plastic foam blocks. *Pad cutters* slice foam rubber blocks to specified thicknesses for such objects as seat cushions and ironing board pads.

Many products undergo further processing to finish them. *Foam-gun operators* reinforce and insulate plastic products such as bathtubs and auto body parts by spraying them with plastic foam. *Plastic-sheet cutters* use power shears to cut sheets, following patterns glued to the sheets by pattern hands. *Sawyers* cut rods, tubes, and sheets to specified dimensions. *Trimmers* trim plastic parts to size using a template and power saw. *Machine finishers* smooth and polish the surface of plastic sheets. And *plastics heat welders* use hot-air guns to fuse together plastic sheets.

Hand finishers trim and smooth products using hand tools and sandpaper. *Buffers* remove ridges and rough edges from fiberglass or plastic castings. Sponge buffers machine-buff the edges of plastic sponges to round them, and *pointing-machine operators* round the points on the teeth of plastic combs.

Edge grinders tend machines that square and smooth edges of plastic floor tile.

Some plastics workers (*assemblers*) and *laminated plastics assemblers-and-gluers* assemble pieces to form certain products. These may include skylights (skylight assemblers) and wet suits (wet suit gluers). Other workers are lacquerers, embossers, printers, carvers, or design inserters. *Plastics inspectors* inspect and test finished products for strength, size, uniformity, and conformity to specifications.

Experienced workers supervise plastics-making departments, and the industry also employs unskilled workers such as laborers to help haul, clean, and assemble plastics materials, equipment, and products.

Requirements

High School

You should take courses in mathematics, chemistry, physics, computer science, shop, drafting, and mechanical drawing. English and speech classes will help build good communications and interpersonal skills.

Postsecondary Training

You'll need a high school diploma to enter the field, and you'll learn most of your skills on the job. In extrusion plants, trainees can become Class I extruders after about three months. Other jobs require training from one to 12 months.

Applicants with some knowledge of chemistry, mathematics, physics, drafting, industrial technology, or computer science have a better chance of being hired. Some colleges offer associate's or bachelor's degrees in plastics technology. Job seekers with these degrees have a definite competitive edge and may also advance more quickly.

Another training option is to participate in an apprenticeship program. Apprenticeships provide experience and a chance to explore the field. Apprenticeships in tool and die making for plastics last four or five years and teach through classroom instruction and on-the-job training. A high school education is normally a prerequisite for an apprenticeship.

Certification or Licensing

Certification isn't required of plastics technicians, but it is available through the Society of the Plastics Industry (SPI). As industry equipment becomes more complex, employers may prefer to hire only certified technicians. To become an NCP Certified Operator, you'll take an exam in one of four areas: blow molding, extrusion, injection molding, or thermoforming. The exam is open to anyone seeking a career in the plastics industry, but you'll likely need at least two years of plastics experience to pass the exam.

Other Requirements

You must have mechanical aptitude and manual dexterity to work well with tools and various materials. Lifting equipment and materials takes some strength, and workers who operate machines stand much of the time. You must be able to work well with others and follow oral and written directions. You must be precise and organized in your work.

"I'm really particular about my work from being in construction for six years," Griggs says. "I pay really close attention to detail."

Exploring

Many high schools are beginning to offer vocational programs, and other apprenticeship opportunities, for those interested in becoming technicians; some of these programs have courses geared specifically toward preparation for the plastics industry. SPI is currently involved in providing career direction to young people interested in the plastics industry. Contact SPI for career and industry information. You can also learn about the industry by reading trade magazines such as *Modern Plastics*. *Plastics News* (http://www.plasticsnews.com) publishes many informative articles on the Web, including rankings of plastics manufacturers.

Employers

Major plastics employers in the United States include DuPont, General Motors, and Owens-Corning. Some of the top thermoforming companies are in Illinois: Tenneco Packaging, Solo Cup Company, and Ivex Packaging Corporation are a few of them. Michigan has some of the top injection molding companies, including Textron, Lear Corporation, UT Automotive, and Venture Industries Corporation. But large plastics companies are located all across the country. According to the SPI, the top plastics industry states ranked by employment are California, Ohio, Michigan, Illinois, and Texas.

Starting Out

After receiving a high school diploma, you should apply directly to the personnel departments of plastics plants in the area in which you wish to work. Newspaper ads may list openings in the industry, and state employment agencies may also provide leads. The Web site http://www.polysort.com features a "virtual job fair" which offers free access to job listings in the plastics industry. The Plastics Molders and Manufacturers Association, a division of the Society of Manufacturing Engineers (http://www.sme.org/), also maintains a job database for members.

Advancement

In the plastics industry, advancement comes with experience, skill, and education. Because plants like to teach workers their own methods, and because skilled plastics workers are scarce, most plastics companies promote workers from within to fill more responsible and higher-paying jobs. Plastics workers who understand machine setup and the properties of plastics will advance more quickly than those limited to machine operations.

Workers who pursue bachelor's or associate's degrees in plastics technology have the best chances for advancement. With advanced training and experience some plastics workers may become plastics engineers or mold designers. Others may move into supervisory, management, or sales and marketing positions. Apprenticeships, such as in tool and die manufacturing, may also lead to more highly paid production work.

Earnings

According to wage surveys conducted by SPI, material handlers in the plastics industry earn about $19,000 to $21,000 a year. In entry-level, or apprenticeship, positions, moldmakers earn between $20,000 and $28,000 a year. Experienced moldmakers earn $31,000 to $40,000 a year.

In addition to salary, many employers offer medical and dental benefits, life insurance, paid sick leave, personal and vacation days, and retirement plans. Employees may also be able to participate in profit-sharing plans.

Work Environment

Most plastics industry workers work 40 hours per week. Because plants operate on 3 shifts, entry-level workers may work nights and move to day shifts as they gain experience and seniority.

Plastics plants are generally safe, well lighted and ventilated, and modern. Workers must observe safety precautions when working around hot machines and plastics, sharp machine parts, and electrical wiring, and when sawing, cutting, or drilling plastics parts. Plastics work, however, is not usually strenuous. Workers use machines to lift heavy dies and other equipment.

As with most production work, jobs in the plastics industry often demand a fair amount of repetition. Workers who need great variety in their jobs may not enjoy production work. Plastics plants tend to be smaller than many other types of factories so a sense of teamwork often develops among the production workers. Such camaraderie can lead to increased job satisfaction and enjoyment.

"There are some safety issues," Marvin said, "working with power tools." But he hasn't encountered too many negatives since taking the job, despite the many hours on his feet, and being restricted to certain areas most of the day. He benefits from a retirement plan and profit-sharing and has an employer who makes an effort to get to know the members of the company.

Outlook

Increased opportunities in foreign markets, the development of new compounds, and increased competition will likely spur the industry as a whole to new economic heights. As a result, the employment of many types of plastics products manufacturing workers is expected to increase faster than the average for all occupations through the year 2006. As more plastics products are substituted for paper, glass, and metal products, more plastics workers will be needed. Molding machine operators, which constitute a large percentage of the workforce, will enjoy a 30 percent increase in growth. Other growth occupations include managers and executives, industrial machine operators, assemblers and fabricators, tool and die makers, extruding and forming machine operators, and cutters and trimmers. Occupations which may experience a decline in employment include grinding-machine operators (due to automation), blenders, and color mixers.

For More Information

For a career brochure and information about education and certification, contact:

Society of the Plastics Industry
1801 K Street, NW, Suite 600K
Washington, DC 20006-1301
Tel: 202-974-5200
Web: http://www.socplas.org

The APC is a trade industry that offers a great deal of information about the plastics industry, and maintains an informative Web site:

American Plastics Council
1801 K Street, NW, Suite 701-L
Washington, DC 20006-1301
Tel: 800-243-5790
Web: http://www.plastics.org

For information about scholarships, seminars, and training, contact:

Plastics Institute of America
University of Massachusetts-Lowell
333 Aiken Street
Lowell, MA 01854
Tel: 978-934-3130
Web: http://www.eng.uml.edu/dept/PIA/index.html

Quality Control Engineers and Technicians

	School Subjects
Mathematics Physics	
	Personal Skills
Mechanical/manipulative Technical/scientific	
	Work Environment
Primarily indoors Primarily one location	
	Minimum Education Level
Associate's degree	
	Salary Range
$17,000 to $40,000 to $70,000	
	Certification or Licensing
Voluntary	
	Outlook
About as fast as the average	

Overview

Quality control engineers plan and direct procedures and activities involved in the processing and production of materials and goods in order to ensure specified standards of quality. They select the best techniques for a specific process or method, determine the level of quality needed, and take the necessary action to maintain or improve quality performance. *Quality control technicians* assist quality control engineers in devising quality control procedures and methods, implement quality control techniques, test and inspect products during different phases of production, and compile and evaluate statistical data to monitor quality levels.

History

Quality control technology is an outgrowth of the Industrial Revolution. As it began in England in the 18th century, each person involved in the manufacturing process was responsible for a particular part of the process. The worker's responsibility was further specialized by the introduction of the concept of interchangeable parts in the late 18th and early 19th centuries. In a manufacturing process using this concept, a worker could concentrate on making just one component, while other workers concentrated on creating other components. Such specialization led to increased production efficiency, especially as manufacturing processes became mechanized during the early part of the 20th century. It also meant, however, that no one worker was responsible for the overall quality of the product. This led to the need for another kind of specialized production worker whose primary responsibility was not one aspect of the product but rather its overall quality.

This responsibility initially belonged to the mechanical engineers and technicians who developed the manufacturing systems, equipment, and procedures. After World War II, however, a new field emerged that was dedicated solely to quality control. Along with specially trained persons to test and inspect products coming off assembly lines, new instruments, equipment, and techniques were developed to measure and monitor specified standards.

At first, quality control engineers and technicians were primarily responsible for random checks of products to ensure they met all specifications. This usually entailed testing and inspecting either finished products or products at various stages of production.

During the 1980s, a quality movement spread across the United States. Faced with increased global competition, especially from Japanese manufacturers, many U.S. companies sought to improve quality and productivity. Quality improvement concepts, such as total quality management, continuous improvement, quality circles, and zero defects gained popularity and changed the way companies viewed quality and quality control practices. A new philosophy emerged, emphasizing quality as the concern of all individuals involved in producing goods and directing that quality be monitored at all stages of manufacturing—not just at the end of production or at random stages of manufacturing.

Today, most companies focus on improving quality during all stages of production, with an emphasis on preventing defects rather than merely identifying defective parts. There is an increased use of sophisticated automated equipment that can test and inspect products as they are manufactured. Automated equipment includes cameras, X rays, lasers, scanners, metal detectors, video inspection systems, electronic sensors, and machine vision systems that can detect the slightest flaw or variance from accepted tolerances. Many companies use statistical process control to record levels of quality and determine the best manufacturing and quality procedures. Quality control engineers and technicians work with employees from all departments of a company to train them in the best quality methods and to seek improvements to manufacturing processes to further improve quality levels.

Many companies today are seeking to conform to international standards for quality, such as ISO 9000, in order to compete with foreign companies and to sell products to companies in countries around the world. These standards are based on concepts of quality regarding industrial goods and services and include documenting quality methods and procedures.

The Job

Quality control engineers are responsible for developing, implementing, and directing processes and practices that result in the desired level of quality for manufactured parts. They identify standards to measure the quality of a part or product, analyze factors that affect quality, and determine the best practices to ensure quality.

Quality control engineers set up procedures to monitor and control quality, devise methods to improve quality, and analyze quality control methods for effectiveness, productivity, and cost factors. They are involved in all aspects of quality during a product's life cycle. Not only do they focus on ensuring quality during production operations, they also get involved in product design and product evaluation. Quality control engineers may be specialists who work with engineers and industrial designers during the design phase of a product, or they may work with sales and marketing professionals to evaluate reports from consumers on how well a product is performing. Quality control engineers are responsible for ensuring that all incoming materials used in a finished product meet required standards and that all instruments and automated equipment used to test and monitor parts during production perform properly. They supervise and direct workers involved in assuring quality, including quality control technicians, inspectors, and related production personnel.

Quality control technicians work with quality control engineers in designing, implementing, and maintaining quality systems. They test and inspect materials and products during all phases of production in order to ensure they meet specified levels of quality. They may test random samples of products or monitor production workers and automated equipment that inspect products during manufacturing. Using engineering blueprints, drawings, and specifications, they measure and inspect parts for dimensions, performance, and mechanical, electrical, and chemical properties. They establish tolerances, or acceptable deviations from engineering specifications, and direct manufacturing personnel in identifying rejects and items that need to be reworked. They monitor production processes to be sure that machinery and equipment are working properly and set to established specifications.

Quality control technicians also record and evaluate test data. Using statistical quality control procedures, technicians prepare charts and write summaries about how well a product conforms to existing standards. Most important, they offer suggestions to quality control engineers on how to modify existing quality standards and manufacturing procedures. This helps to achieve the optimum product quality from existing or proposed new equipment.

Quality control technicians may specialize in any of the following areas: product design, incoming materials, process control, product evaluation, inventory control, product reliability, research and development, and administrative applications. Nearly all industries employ quality control technicians.

Requirements

High School

In high school, prospective engineers and technicians should take classes in English, mathematics (including algebra and geometry), physical sciences, physics, and chemistry. They should also take shop, mechanical drawing, and computer courses. Students should especially seek English courses that will develop their reading skills, the ability to write short reports with good organization and logical development of ideas, and the ability to speak comfortably and effectively in front of a group.

Postsecondary Training

Quality control engineers must have a bachelor's degree in engineering. Many quality control engineers receive degrees in industrial or manufacturing engineering. Some receive degrees in metallurgical, mechanical, electrical, or chemical engineering depending on where they plan to work. College engineering programs vary based on the type of engineering program. Most programs take four to five years to complete and include courses in mathematics, physics, and chemistry. Other useful courses include statistics, logistics, business management, and technical writing.

Educational requirements for quality control technicians vary by industry. Most employers of quality control technicians prefer to hire applicants who have received some specialized training. A small number of positions for technicians require a bachelor of arts or science degree. In most cases, though, completion of a two-year technical program is sufficient. Students enrolled in such a program at a community college or technical school take courses in the physical sciences, mathematics, materials control, materials testing, and engineering-related subjects.

Certification or Licensing

Although there are no licensing or certification requirements designed specifically for quality control engineers or technicians, some may need to meet special requirements that apply only within the industry employing them. Many quality control engineers and technicians pursue voluntary certification to indicate that they have achieved a certain level of competency, either through education or work experience. Such certification is offered through professional associations, such as the American Society for Quality Control (ASQC), and requires passing an examination. Many employers value this certification and regard it as a demonstration of professionalism.

Other Requirements

Quality control engineers need scientific and mathematical aptitudes, strong interpersonal skills, and leadership abilities. Good judgment is also needed, as quality control engineers must weigh all the factors influencing quality and determine procedures that incorporate price, performance, and cost factors.

Quality control technicians should enjoy and do well in mathematics, science, and other technical subjects and should feel comfortable using the language and symbols of mathematics and science. They should have good eyesight and good manual skills, including the ability to use hand tools. They should be able to follow technical instructions and to make sound judgments about technical matters. Finally, they should have orderly minds and be able to maintain records, conduct inventories, and estimate quantities.

Exploring

Because quality control engineers and technicians work in a wide variety of settings, prospective engineers and technicians who want to learn more about quality control technology can consider a range of possibilities for experiencing or further exploring such work. Quality control activities are often directly involved with manufacturing processes. Students may be able to get part-time or summer jobs in manufacturing settings, even if not specifically in the quality control area. Although this type of work may consist of menial tasks, it does offer firsthand experience and demonstrates interest to future employers.

Quality control engineers and technicians work with scientific instruments; therefore, academic or industrial arts courses that introduce different kinds of scientific or technical equipment will be helpful, along with electrical and machine shop courses, mechanical drawing courses, and chemistry courses with lab sections. Joining a radio, computer, or science club is also a good way to gain experience and to engage in team-building and problem-solving activities. Active participation in clubs is a good way to learn skills that will benefit you when working with other professionals in manufacturing and industrial settings.

Employers

The majority of quality control engineers and technicians are employed in the manufacturing sector of the economy. Because engineers and technicians work in all areas of industry, their employers vary widely in size, product, location, and prestige.

Starting Out

Students enrolled in two-year technical schools may learn of openings for quality control technicians through their schools' job placement services. Recruiters often visit these schools and interview graduating students for technical positions. Quality control engineers also may learn of job openings through their schools' job placement services, recruiters, and job fairs. In many cases, employers prefer to hire engineers who have some work experience in their particular industry. For this reason, applicants who have had summer or part-time employment or participated in a work-study or internship program have greater job opportunities.

Students may also learn about openings through help wanted ads or by using the services of state and private employment services. They also may apply directly to companies that employ quality control engineers and technicians. Students can identify and research such companies by using job resource guides and other reference materials available at most public libraries.

Advancement

Quality control technicians usually begin their work under the direct and constant supervision of an experienced technician or engineer. As they gain experience or additional education, they are given more responsible assignments. They can also become quality control engineers with additional education. Promotion usually depends upon additional training as well as job performance. Technicians who obtain additional training have greater chances for advancement opportunities.

Quality control engineers may have limited opportunities to advance within their companies. However, because quality control engineers work in all areas of industry, they have the opportunity to change jobs or companies to pursue more challenging or higher-paying positions. Quality control engineers who work in companies with large staffs of quality personnel can become quality control directors or advance to operations management positions.

Earnings

Earnings vary according to the type of work, the industry, and the geographical location. Quality control engineers earn salaries comparable to other engineers. Beginning engineers with a bachelor's degree generally earn between $31,000 and $35,000 a year. Those with master's degrees earn salaries of about $41,322 in their first jobs upon graduation. Experienced quality control engineers earn salaries ranging from $35,000 to $70,000.

Most beginning quality control technicians who are graduates of two-year technical programs earn salaries ranging from $17,000 to $21,000 a year. Experienced technicians with two-year degrees earn salaries that range from $21,000 to $36,000 a year; some senior technicians with special skills or experience may earn much more.

Most companies offer benefits that include paid vacations, paid holidays, and health insurance. Actual benefits depend upon the company, but may also include pension plans, profit sharing, 401(k) plans, and tuition assistance programs.

Work Environment

Quality control engineers and technicians work in a variety of settings, and their conditions of work vary accordingly. Most work in manufacturing plants, though the type of industry determines the actual environment. For example, quality control engineers in the metals industry usually work in foundries or iron and steel plants. Conditions are hot, dirty, and noisy. Other factories, such as for the electronics or pharmaceutical industries, are generally quiet and clean. Most engineers and technicians have offices separate from the production floor, but they still need to spend a fair amount of time there. Engineers and technicians involved with testing and product analysis work in comfortable surroundings, such as a laboratory or workshop. Even in these settings, however, they may be exposed to unpleasant fumes and toxic chemicals. In general, quality control engineers and technicians work inside and are expected to do some light lifting and carrying (usually not more than 20 pounds). Because many manufacturing plants operate 24 hours a day, some quality control technicians may need to work second or third shifts.

As with most engineering and technical positions, the work can be both challenging and routine. Engineers and technicians can expect to find some tasks repetitious and tedious. In most cases, though, the work provides variety and satisfaction from using highly developed skills and technical expertise.

Outlook

The employment outlook depends, to some degree, on general economic conditions. Although many economists forecast low to moderate growth in manufacturing operations through the year 2007, employment opportunities for quality control personnel should remain steady or slightly increase as many companies place increased emphasis on quality control activities.

Many companies are making vigorous efforts to make their manufacturing processes more efficient, lower costs, and improve productivity and quality. Opportunities for quality control engineers and technicians should be good in the food and beverage industries, pharmaceutical firms, electronics companies, and chemical companies. Quality control engineers and technicians also may find employment in industries using robotics equipment or in the aerospace, biomedical, bioengineering, environmental controls, and transportation industries. Lowered rates of manufacturing in the automotive and defense industries will decrease the number of quality control personnel needed for these areas. Declines in employment in some industries may occur because of the increased use of automated equipment that tests and inspects parts during production operations.

For More Information

American Society for Quality Control
PO Box 3005
Milwaukee, WI 53201-3005
Tel: 800-248-1946

Textile Manufacturing Workers

Computer science **Technical/Shop**	School Subjects
Following instructions **Mechanical/manipulative**	Personal Skills
Primarily indoors **Primarily one location**	Work Environment
High school diploma	Minimum Education Level
$22,000 to $27,000 to $35,000	Salary Range
None available	Certification or Licensing
Decline	Outlook

Overview

Workers in textile manufacturing occupations are concerned with preparing natural and synthetic fibers for spinning into yarn and manufacturing yarn into textile products that are used in clothing, household goods, and for many industrial purposes. Among the processes that these workers perform are cleaning, carding, combing, and spinning fibers; weaving, knitting, or bonding yarns and threads into textiles; and dyeing and finishing fabrics.

History

Archaeological evidence suggests that people have been weaving natural fibers into cloth for at least 7,000 years. Basketweaving probably preceded and inspired the weaving of cloth. By about 5,000 years ago, cotton, silk, linen, and wool fabrics were being produced in several areas of the world. While ancient weavers used procedures and equipment that seem simple by today's standards, some of the cloth they made was of fine quality and striking beauty.

Over time, the production of textiles grew into a highly developed craft industry with various regional centers that were renowned for different kinds of textile products. Yet, until the 18th century, the making of fabrics was largely a cottage industry in which no more than a few people, often family groups, worked in small shops with their own equipment to make products by hand. With the Industrial Revolution and the invention of machines such as the cotton gin and the power loom, a wide variety of textiles could be produced in factories at low cost and in large quantities. Improvements have continued into the 20th century, so that today many processes in making textiles are highly automated.

Other changes have revolutionized the production of fabrics. The first attempts to make artificial fibers date to the 17th century, but it was not until the late 19th and early 20th centuries that a reasonably successful synthetic, a kind of rayon, was developed from the plant substance cellulose. Since then, hundreds of synthetic fibers have been developed from such sources as coal, wood, ammonia, and proteins. Other applications of science and technology to the textile industry have resulted in cloth that has a wide variety of attractive or useful qualities. Many fabrics that resist creases, repel stains, or are fireproof, mothproof, antiseptic, nonshrinking, glazed, softened, or stiff are the product of modern mechanical or chemical finishing.

Of the textiles produced in the United States today, only about half are used for wearing apparel. The rest are used in household products (towels, sheets, upholstery) and industrial products (conveyor belts, tire cords, parachutes).

The Job

Most textile workers operate or tend machines. In the most modern plants, the machines are often quite sophisticated and include computerized controls.

Workers in textile manufacturing can be grouped in several categories. Some workers operate machines that clean and align fibers, draw and spin them into yarn, and knit, weave, or tuft the yarn into textile products. Other workers, usually employees of chemical companies, tend machines that produce synthetic fibers through chemical processes. Still other workers prepare machines before production runs. They set up the equipment, adjusting timing and control mechanisms, and often maintain the machines as well. Another category of workers specializes in finishing textile products before they are sent out to consumers. The following paragraphs describe just a few of the many kinds of specialized workers in textile manufacturing occupations.

In the transformation of raw fiber into cloth, one of the first steps may be performed by *staple cutters*. They place opened bales of raw stock or cans of sliver (combed, untwisted strands of fiber) at the feed end of a cutting machine. They guide the raw stock or sliver onto a conveyor belt or feed rolls, which pull it against the cutting blades. They examine the cut fibers as they fall from the blades and measure them to make sure they are the required length.

Spinneret operators oversee machinery that makes manufactured fibers from such nonfibrous materials as metal or plastic. Chemical compounds are dissolved or melted in a liquid, which is then extruded, or forced, through holes in a metal plate, called a spinneret. The size and shape of the holes determine the shape and uses of the fiber. Workers adjust the flow of fiber base through the spinneret, repair breaks in the fiber, and make minor adjustments to the machinery.

Frame spinners, also called *spinning-frame tenders,* tend machines that draw out and twist the sliver into yarn. These workers patrol the spinning-machine area to ensure that the machines have a continuous supply of sliver or roving (a soft, slightly twisted strand of fiber made from sliver). They replace nearly empty packages of roving or sliver with full ones. If they detect a break in the yarn being spun, or in the roving or sliver being fed into the spinning frame, they stop the machine and repair the break. They are responsible for keeping a continuous length of material threaded through the spinning frame while the machine is operating.

Spinning supervisors supervise and coordinate the activities of the various spinning workers. From the production schedule, they determine the quantity and texture of yarn to be spun and the type of fiber to be used. Then they compute such factors as the proper spacing of rollers and the correct size of twist gears, using mathematical formulas and tables and their knowledge of spinning machine processes. As the spun yarn leaves the spinning frame, they examine it to detect variations from standards.

A *textile production worker* adjusts the tension on one of the rapier weaving machines. Once the fiber is spun into yarn or thread, it is ready for weaving, knitting, or tufting. Woven fabrics are made on looms that interlace the threads. Knit products, such as socks or women's hosiery, are produced by intermeshing loops of yarn. The tufting process, used in making carpets, involves pushing loops of yarn through a material backing.

Beam-warper tenders work at high-speed warpers, which are machines that automatically wind yarn onto beams, or cylinders, preparatory to dyeing or weaving. A creel, or rack of yarn spools, is positioned at the feed end of the machine. The workers examine the creel to make sure that the size, color, number, and arrangement of the yarn spools correspond to specifications. They thread the machine with the yarn from the spools, pulling the yarn through several sensing devices, and fastening the yarn to the empty cylinder. After setting a counter to record the amount of yarn wound, they start the machine. If a strand of yarn breaks, the machine stops, and the tenders locate and tie the broken ends. When the specified amount of yarn has been wound, they stop the machine, cut the yarn strands, and tape the cut ends.

Weavers or *loom operators* operate a battery of automatic looms that weave yarn into cloth. They observe the cloth being woven carefully to detect any flaws, and they remove weaving defects by cutting out the filling (cross) threads in the area. If a loom stops, they locate the problem and either correct it or, in the case of mechanical breakdown, notify the appropriate repairer.

After the fabric is removed from the loom, it is ready for dyeing and finishing, which includes treating fabrics to make them fire-, shrink-, wrinkle-, or soil-resistant.

Dye-range operators control the feed end of a dye range, which is an arrangement of equipment that dyes and dries cloth. Operators position the cloth to be dyed and machine-sew its end to the end of the cloth already in the machine. They turn valves to admit dye from a mixing tank onto the dye pads, and they regulate the temperature of the dye and the air in the drying box. They start the machine, and when the process is complete, they record yardage dyed, lot numbers, and the machine running time. *Colorists, screen printing artists, screen makers,* and *screen printers* print designs on textiles.

Cloth testers perform tests on gray goods and finished cloth samples. They may count the number of threads in a sample, test its tensile strength in a tearing machine, and crease it to determine its resilience. They may also test for such characteristics as abrasion resistance, fastness of dye, flame retardance, and absorbency, depending on the type of cloth.

Requirements

High School

For some textile production jobs, a high school education is desirable but may not be necessary. Workers who operate machines are often hired as unskilled labor and trained on the job. However, with the increasingly complex machinery and manufacturing methods in this industry, more and more often a high school diploma plus some technical training is expected of job applicants. High school students interested in a textile career should take courses in physics, chemistry, mathematics, and English. Computer skills are necessary, since many machines are now operated by computer technology.

Postsecondary Training

Even those with postsecondary school education generally must go through a period of on-the-job training by experienced workers or representatives of equipment manufacturers where they learn the procedures and systems of their particular company. Some companies have coop programs with nearby schools. Participants in these programs work as interns during their academic training with the agreement that they will work for the sponsoring company upon graduation. A two-year associate degree in textile technology is required for technicians, laboratory testers, and supervisory personnel.

Other Requirements

Many machine operators need physical stamina, manual dexterity, and a mechanical aptitude to do their job. Changes are under way in the industry that make other kinds of personal characteristics increasingly important, such as the ability to assume responsibility, to take initiative, to communicate with others, and to work well as a part of a team.

About 15 percent of all textile production workers belong to a union, such as the Union of Needle Trades, Industrial, and Textile Employees (UNITE).

Exploring

High school courses in subjects such as shop, mechanical drawing, and chemistry and hobbies involving model-building and working with machinery can be good preparation for many jobs in the textile manufacturing field. Students may be able to find summer employment in a textile plant. If that cannot be arranged, a machine operator's job in another manufacturing industry may provide a similar enough experience that it is useful in understanding something about textile manufacturing work.

Employers

Most textile production workers are employed either in mills that spin and weave "gray goods," meaning raw, undyed, unfinished fabrics, or in finishing plants, where gray goods are treated with processes like dyeing and bleaching. Some textile companies combine these two stages of manufacturing under one roof.

Employment opportunities for textile manufacturing workers are concentrated in the South and the Northeast. Over half of the jobs in this industry are located in the states of North Carolina, Georgia, and South Carolina.

Starting Out

Most textile production workers obtain their jobs by answering newspaper advertisements or by applying directly to the personnel office of a textile plant. A new worker usually receives between a week and several months of on-the-job training, depending on the complexity of the job.

Graduates of textile technology programs in colleges and technical institutes may be informed about job openings through their school's placement office. They may be able to line up permanent positions before graduation. Sometimes students in technical programs are sponsored by a local textile company, and upon graduation, they go to work for the sponsoring company.

Advancement

Production workers in textile manufacturing who become skilled machine operators may be promoted to positions in which they train new employees. Other workers can qualify for better jobs by learning additional machine-operating skills. Usually the workers with the best knowledge of machine operations are those who set up and prepare machines before production runs. Skilled workers who show that they have good judgment and leadership abilities may be promoted to supervisory positions, in charge of a bank of machines or a stage in the production process. Some companies offer continuing education opportunities to dedicated workers.

Laboratory workers may advance to supervisory positions in the lab. If their educational background includes such courses as industrial engineering and quality control, they may move up to management jobs where they plan and control production.

Earnings

Earnings of textile industry workers vary depending on the type of plant where they are employed and the workers' job responsibilities, the shift they work, and seniority. Overall, the average annual wage for textile workers in 1998 was $425 a week or about $22,000 a year, according to the American Textile Manufacturers Institute. Workers at plants located in the North tend to be paid more than those in the South.

Beginning laboratory testers and technicians with associate degrees in textile technology can earn annual salaries ranging from $27,000 to about $35,000 after a few years of experience, according to the North Carolina Center for Applied Textile Technology. Salaries generally increase with more education and greater responsibility.

Most workers with a year or more of service receive paid vacations and insurance benefits. Many are able to participate in pension plans, profit sharing, or year-end bonuses. Some companies offer their employees discounts on the textiles or textile products they sell.

Work Environment

Work areas in modern textile plants are largely clean, well lighted, air-conditioned, and humidity-controlled. Older facilities may be less comfortable, with more fibers or fumes in the air, requiring some workers to wear protective glasses or masks. Some machines can be very noisy, and workers near them must wear ear protectors. Workers also must stay alert and use caution when working around high-speed machines that can catch clothing or jewelry. Those who work around chemicals must wear protective clothing and sometimes respirators. Increasing attention to worker safety and health has forced textile manufacturing companies to comply with tough federal, state, and local regulations.

Workweeks in this industry average 40 hours in length. Depending on business conditions, some plants may operate 24 hours a day, with three shifts a day. Production employees may work rotating shifts, so that they share night and weekend hours. Some companies have a four-shift continuous operating schedule, consisting of a 168-hour workweek made up of four daily shifts totaling 42 hours a week. This system offers a rotating arrangement of days off. During production cutbacks, companies may go to a three- or four-day workweek, but they generally try to avoid layoffs during slow seasons.

Machine operators are often on their feet during much of their shift. Some jobs involve repetitive tasks that some people find boring.

Outlook

In 1997 more than 600,000 people were employed in the U.S. textile industry. About 500,000 were machine operators in textile mills. Over the next 10 to 15 years, employment in this field is expected to decline, even as the demand for textile products increases.

Changes in the textile industry will account for much of this decline. Factories are reorganizing production operations for greater efficiency and installing equipment that relies on more highly automated and computerized machines and processes. Such technology as shuttleless and air-jet looms and computer-controlled machinery allows several machines to be operated by one operator and still increase speed and productivity.

Another factor that will probably contribute to a reduced demand for U.S. textile workers is an increase in imports of textiles from other countries. There is a continuing trend toward freer world markets and looser trade restrictions.

While fewer workers will be needed to operate machines, there will continue to be many job openings each year as experienced people transfer to other jobs or leave the workforce. Workers who have good technical training and skills will have the best job opportunities. In fact, the demand for highly skilled workers, such as scientists, engineers, and computer specialists, will be even greater as companies strive to expand their markets globally.

For More Information

This union fights for workers' rights and represents workers in various industries, including basic apparel and textiles. It was recently formed from a merger of International Ladies' Garment Workers Union and the Amalgamated Clothing and Textile Workers Union.

UNITE (Union of Needle Trades, Industrial and Textile Employees)
1710 Broadway
New York, NY 10019
Tel: 212-265-7000
Web: http://www.uniteunion.org

This national trade association for the U.S. textile industry has member companies that operate in more than 30 states and process about 80 percent of all textile fibers consumed by plants in the United States. It works to encourage global competitiveness and increase foreign market access.

American Textile Manufacturers Institute
1130 Connecticut Avenue, NW, Suite 1200
Washington, DC 20006
Tel: 202-862-0500
Web: http://www.atmi.org

The following research and graduate education organization offers a master's degree in textile technology. It also offers fee-based library and information services.

Institute of Textile Technology
2551 Ivy Road
Charlottesville, VA 22903
Tel: 804-296-5511
Web: http://www.itt.edu

Tobacco Products Industry Workers

Agriculture Biology	School Subjects
Following instructions Mechanical/manipulative	Personal Skills
Primarily indoors Primarily one location	Work Environment
High school diploma	Minimum Education Level
$25,000 to $39,600 to $50,700	Salary Range
None available	Certification or Licensing
Decline	Outlook

Overview

Tobacco products industry workers manufacture cigars, cigarettes, chewing tobacco, smoking tobacco, and snuff from leaf tobacco. They dry, cure, age, cut, roll, form, and package tobacco in products used by millions of people in the United States and in other countries around the world.

History

The use of tobacco has been traced back to Mayan cultures of nearly 2,000 years ago. As the Mayas moved north, through Central America and into North America, tobacco use spread throughout the continent. When Christopher Columbus arrived in the Caribbean in 1492, he was introduced to tobacco smoking by the Arawak tribe, who smoked the leaves of the plant rolled into cigars. Tobacco seeds were brought back to Europe, where they were cultivated. The Europeans, believing tobacco had medicinal properties, quickly adopted the practice of smoking. Sir Walter Raleigh popularized pipe smoking around 1586, and soon the growing and use of tobacco spread around the world.

Tobacco growing became an important economic activity in America beginning in the colonial era, in part because of the ideal growing conditions found in many of the Southern and Southeastern colonies. Tobacco quickly became a vital part of the colonies' international trade.

Tobacco use remained largely limited to small per-person quantities until the development of cigarettes in the mid-1800s. The invention of the cigarette-making machine in 1881 made the mass production of cigarettes possible. Nevertheless, the average person smoked only 40 cigarettes per year. It was only in the early decades of the 20th century that cigarette consumption, spurred by advertising campaigns, became popular across the country. Soon, the average person smoked up to 40 cigarettes per day.

By the 1960s, it became increasingly apparent that tobacco use was detrimental to people's health. In 1969, laws were passed requiring warning labels to be placed on all tobacco products. During the 1970s, increasing agitation by the antismoking movement led to laws, taxes, and other regulations being placed on the sale and use of tobacco products. Many other countries followed with similar laws and regulations. The number of smokers dropped by as much as 30 percent, and those who still smoked, smoked less. In response, the tobacco industry introduced products such as light cigarettes and low tar and low nicotine cigarettes. In the late 1990s, the tobacco industry was at the center of debate, controversy, and subsequent state lawsuits over addictive substances and cancer-causing agents contained in cigarettes. This controversy and the declining numbers of smokers in the United States and much of the West have had a strong impact on the employment levels in the tobacco industry.

The Job

Various kinds of tobacco plants are cultivated for use in tobacco products. After harvesting, the different types of tobacco are processed in different ways. Using one method or another, all tobacco is cured, or dried, for several days to a month or more in order to change its physical and chemical characteristics. Farmers sometimes air-cure tobacco by hanging it in barns to dry naturally. Other curing methods are fire-curing in barns with open fires and flue-curing in barns with flues that circulate heat. Some tobacco is sun-cured by drying it outdoors in the sun.

Cured tobacco is auctioned to tobacco product manufacturers or other dealers. The first step in the manufacturing process is separating out stems, midribs of leaves, and foreign matter. Usually this is done by workers who feed the tobacco into machines. Once stemmed, the tobacco is dried again by redrying-machine operators, who use machines with hot-air blowers and fans.

The tobacco is then packed for aging. In preparation for packing, workers may adjust the moisture content of the dry tobacco by steaming the leaves or wetting them down with water. The tobacco is prized, or packed, into large barrels or cases that can hold about a thousand pounds of tobacco each. Workers, including *bulkers, prizers,* and *hydraulic-press operators,* pack the containers, which go to warehouses to be aged. The aging process, which may take up to two years, alters the aroma and flavor of the tobacco. After it is aged, workers take the tobacco to factories, where it is removed from the containers.

The tobacco is further conditioned by adding moisture. *Blenders* then select tobacco of various grades and kinds to produce blends with specific characteristics or for specific products, such as cigars or snuff. They place the tobacco on conveyors headed for processing. Blending laborers replenish supplies of the different tobaccos for the blending line. Blending-line attendants tend the conveyors and machines that mix the specified blends.

Some tobacco is flavored using casing fluids, which are water-soluble mixtures. *Casing-material weighers, casing-machine operators, wringer operators, casing cookers,* and *casing-fluid tenders* participate in this flavoring process by preparing the casing material, saturating the tobacco with it, and removing excess fluid before further processing.

The tobacco is ready to be cut into pieces of the correct size. Tobacco for cigars and cigarettes is shredded and cleaned in machines operated by machine filler shredders and strip-cutting-machine operators. *Snuff grinders* and *snuff screeners* tend machines that pulverize chopped tobacco into snuff and sift it through screens to remove oversized particles. *Riddler operators* tend screening devices that separate coarse pieces of tobacco from cut tobacco.

Once cut, the tobacco is made into salable products. Cigarettes are made by machines that wrap shredded tobacco and filters with papers. Various workers feed these machines, make the filters, and run the machines, which also print the company's name and insignia on the rolling papers.

Cigar making is similar, except that the filler tobacco is wrapped in tobacco leaf instead of paper. The filler is held together and formed into a bunch in a binder leaf, and the bunch is rolled in a spiral in a wrapper leaf. Various workers sort and count appropriate wrapper leaves and binder leaves. They roll filler tobacco and binder leaves into bunches by hand or using machines. The bunches are pressed into cigar-shaped molds, and *bunch trimmers* trim excess tobacco from the molds before the bunches are wrapped.

Other workers operate machines that automatically form and wrap cigars. They include auto rollers and wrapper layers, who wrap bunches with sheet tobacco or wrapper leaves. Some workers wrap bunches by hand. *Cigarhead piercers* use machines to pierce draft holes in the cigar ends. Some cigars are pressed into a square shape by tray fillers and press-machine feeders before they are packaged in cigar bands and cellophane. *Patch workers* repair defective or damaged cigars by patching holes with pieces of wrapper leaf.

Some tobacco is made into other products, such as plugs, lumps, and twists. These products are chewed instead of smoked. Twists and some plugs may be made by hand, while most plugs and lumps are made by machine.

Many workers are employed in packaging the manufactured tobacco products. Finally, *tobacco inspectors* check that the products and their packaging meet quality standards, removing items that are defective. The industry also employs a variety of workers to maintain equipment; load, unload, and distribute materials; prepare tobacco for the different stages of processing; salvage defective items for reclamation; and maintain records of tobacco bought and sold.

Requirements

High School

The minimum requirement for all tobacco workers is a high school diploma. Maintenance and mechanical workers often need to be high school graduates with machine maintenance skills or experience. They may need to learn additional skills on the job.

Other Requirements

Workers such as tobacco buyers or graders who must judge tobacco based on its smell, feel, and appearance usually need at least several years' experience working with tobacco to become familiar with its characteristics. Most tobacco products industry workers are members of the Bakery, Confectionery and Tobacco Workers' International Union.

Exploring

Part-time or seasonal tobacco processing jobs may be available for people who are interested in this field. Some plants where tobacco products are manufactured may allow visitors to observe their operations.

Employers

Most jobs in this industry are located in factories close to tobacco-growing regions, especially in the South and Southeast. Most cigarette factories are in North Carolina and Virginia. Many cigar factories are in Florida and Pennsylvania.

Starting Out

Job seekers should apply in person at local tobacco products factories that may be hiring new workers. Leads for specific job openings may be located through the local offices of the state employment service and through union locals. Newspaper classified ads may also carry listings of available jobs.

Advancement

In the tobacco products industry, advancement is related to increased skills. Machine operators may advance by learning how to run more complex equipment. Experienced workers may be promoted to supervisory positions. With sufficient knowledge and experience, some production workers may eventually become tobacco buyers for manufacturers or tobacco graders with the U.S. Department of Agriculture.

Earnings

Wages for tobacco production workers are generally higher than for most other producers of consumable goods. Earnings vary considerably with the plant and the workers' job skills and responsibilities. The average salary for all tobacco production workers in 1995 was around $39,600 per year. Cigarette workers have among the highest earnings of tobacco products workers, earning an average of around $50,700 a year in 1995. Cigar workers and other tobacco products workers tend to earn less. Starting salaries average around $25,000 per year. Tobacco products workers usually receive benefits that include health and life insurance, paid holiday and vacation days, profit-sharing plans, pension plans, and various disability benefits.

Work Environment

In most plants, worker comfort and efficiency are important concerns. Work areas are usually clean, well lighted, and pleasantly air-conditioned. Manufacturing processes are automated wherever possible, and the equipment is designed with safety and comfort in mind. On the downside, much of the work is highly repetitive, and people can find their work very monotonous. Also, tobacco has a strong smell that bothers some people. Some stages of processing produce large quantities of tobacco dust.

Outlook

Employment in the tobacco industry has decreased in recent decades, so that by 1995 there were only about 32,000 workers employed in tobacco production jobs in the United States. About 10,000 tobacco production jobs were lost between 1994 and 1995 alone. This decline is mainly the result of increased automation in manufacturing processes. While Americans are generally using less tobacco, exports of American-made tobacco products are increasing, especially to the former Soviet Union and Eastern Europe, the Middle East, and Asia. Further decline will result from the controversies and lawsuits in this country, which will cost tobacco companies millions of dollars. Most future demand for workers in this industry will probably be because of a need to replace workers who have moved to other jobs or left the workforce entirely.

For More Information

Tobacco Institute
1875 Eye Street, NW, Suite 800
Washington, DC 20006
Tel: 202-457-4800

Toy Industry Workers

	School Subjects
Mathematics Technical/Shop	

	Personal Skills
Technical/scientific Mechanical/manipulative	

	Work Environment
Primarily indoors Primarily one location	

	Minimum Education Level
High school diploma	

	Salary Range
$4.25 per hour to $25,000 to $150,000	

	Certification or Licensing
None available	

	Outlook
About as fast as the average	

Overview

Toy industry workers are the individuals who create, design, manufacture, and market toys and games to adults and children. Their jobs are similar to those of their counterparts in other industries. Some work on large machines, while others assemble toys by hand. According to the U.S. Bureau of Labor Statistics, 39,000 individuals were employed in the toy industry in 1997, approximately 67 percent working in production. Most toy companies are located in or near large metropolitan areas.

History

Toys and games probably have existed as long as there have been humans. Recreational games have roots in ancient cultures. For example, backgammon, one of the oldest known board games, dates back about 5,000 years to areas around the Mediterranean. Chess developed in about the sixth century in India or China and was based on other ancient games.

Dolls and figurines also have turned up among old artifacts. Some seem to have been used as playthings, while others apparently had religious or symbolic importance. More recently, European kings and noblemen gave elaborate dolls in fancy costumes as gifts. Fashion styles thus were spread through other regions and countries. Doll makers in cities such as Paris, France, and Nuremberg, Germany, became famous for crafting especially beautiful dolls. Over the years dolls have been made of wood, clay, china, papier-mâché, wax, and hard rubber, and they have been collected and admired by adults as well as children.

For centuries, most toys were made by hand at home. Mass production began in the 19th century during the Industrial Revolution. In the 20th century, one of the most enduringly popular toys was the teddy bear, named after President Theodore Roosevelt.

Toy companies generally devise their own products or adapt them from perennial favorites, but they occasionally buy ideas for new toys and games from outsiders. One famous example of this was a board game devised during the Depression by an out-of-work man in his kitchen. He drew a playing board on his tablecloth using the names of streets in his hometown of Atlantic City and devised a game that let him act out his fantasies of being a real estate and business tycoon. The game, which he called "Monopoly," became one of the most popular games of all time.

The popularity of certain toys rises and falls over time. Some toys maintain their popularity with successive generations of children or experience a comeback after a few years. Computer and video games have boomed during the past decade and will undoubtedly continue to become more complex and realistic as technology advances. Still, it is very difficult to predict which new toys will become popular. Introducing a new toy into the marketplace is a gamble, and that adds excitement and pressure to the industry.

The Job

Taking a toy from the idea stage to the store shelf is a long and complex operation, sometimes requiring a year or two or even longer. Ideas for new toys or games may come from a variety of sources. In large companies, the marketing department and the research and development department review the types of toys that are currently selling well, and they devise new toys to meet the perceived demand. Companies also get ideas from professional inventors, freelance designers, and ordinary people, including children, who write to them describing new toys they would like to see made.

Toy companies consider ideas for production that they sometimes end up scrapping. A toy company has two main considerations in deciding whether to produce a toy: the degree of interest children (or adults) might have in playing with the toy and whether the company can manufacture it profitably.

A toy must be fun to play with, but there are measures of a toy's worth other than amusement. Some toys are designed to be educational, develop motor skills, excite imagination and curiosity about the world, or help children learn ways of expressing themselves.

Often manufacturers test new ideas to determine their appeal to children. *Model makers* create prototypes of new toys. *Marketing researchers* in the company coordinate sessions during which groups of children play with the prototype toys. If the children in the test group enjoy a toy and return to play with it more than a few times, the toy has passed a major milestone.

The company also has to ask other important questions: Is the toy safe and durable? Is it similar to other toys on the market? Is there potential for a large number of buyers? Can the toy be mass-produced at a low enough cost per toy to ensure a profit? Such questions are usually the responsibility of *research and development workers,* who draw up detailed designs for new toys, determine materials to be used, and devise methods to manufacture the toy economically. After the research and development employees have completed their work, the project is passed on to engineers who start production.

Electronic toys, video games, and computer games have skyrocketed in popularity in the past decade. The people who develop them include *computer engineers, technicians,* and *software programmers. Technical development engineers* work on toys that involve advanced mechanical or acoustical technology. *Plastics engineers* work on plans for plastic toys. They design tools and molds for making plastic toy parts, and they determine the type of molding process and plastic that are best for the job. Plastics engineers who work for large firms may design and build 150 or more new molds each year.

To determine the best way to manufacture a toy, *manufacturing engineers* study the blueprints for the new product and identify necessary machinery. They may decide that the company can modify equipment it already has, or they may recommend purchasing new machinery. Throughout the engineering process, it is important to find ways to minimize production costs while still maintaining quality.

After selecting the equipment for production, *industrial engineers* design the operations of manufacturing: the layout of the plant, the time each step in the process should take, the number of workers needed, the ways to measure performance, and other detailed factors. Next, the engineers teach supervisors and assembly workers how to operate the machinery and assemble the new toy. They inform shift supervisors about the rate of production the company expects. Industrial engineers also might be responsible for designing the process of packaging and shipping the completed toys.

As toys are being built on the assembly line, *quality control engineers* inspect them for safety and durability. Most toy companies adhere to the quality standards outlined in ASTM F963-95, a set of voluntary guidelines the toy industry has developed for itself. The toy industry is also monitored by the Consumer Products Safety Commission and must adhere to various federal laws and standards that cover the safety of toys under normal use and any foreseeable misuse or abuse.

Finally, getting the toys from the factory to the store shelf is the responsibility of *sales and merchandising workers*. These employees stay in contact with toy stores and retail outlets and arrange for toy displays and in-store product promotions.

Factory workers on assembly lines mass-produce practically all toys and games. The manufacturing processes can be as unique as the toys themselves. Workers first cast pieces of plastic toys in injection molds and then assemble them. They machine, assemble, and finish or paint wooden and metal toys. They make board games employing many of the same printing and binding processes used for books. They print the playing surface on a piece of paper, glue it to a piece of cardboard of the proper size, and tape the two halves of the board together with bookbinding equipment.

Toy assemblers put together various plastic, wood, metal, or fabric pieces to complete toys. They may sit at a conveyor belt or workbench, where they use small power tools or hand tools, such as pliers and hammers, to fasten the pieces together. Other toy assemblers operate larger machines such as drill presses, reamers, flanging presses, and punch presses. On toys like wagons that are made on assembly lines, assemblers may do only a single task, such as attaching axles or tires. Other toys may be assembled entirely by one person; for instance, one person at one station on an assembly line may attach the heads, arms, and legs of action figures.

Requirements

High School

Employers usually prefer to hire high school graduates for production jobs in routine assembly of toys and games. Experience in working with machinery, even in high school shop classes, can be a plus. New employees learn most assembly skills on the job from experienced workers over a period of a few days to a few weeks.

Postsecondary Training

Workers need different skills for different types of positions. Engineers and management personnel usually need college degrees. Because of the wide range of activities in the toy business, people may be hired with training in various fields, including art, electronics, architecture, psychology, business, and the sciences.

Certification or Licensing

Many workers in toy factories join a labor union, the Amalgamated, Industrial and Toy and Novelty Workers of America.

Other Requirements

Some people succeed with skills other than those learned in college classrooms. Sales and inventing new toys are among the areas where a worker's success may depend more on creative, innovative approaches than on college training. Some of the production jobs involve repetitive work that must be completed quickly and accurately. Attention to detail is an important quality.

Exploring

To gain some experience working in the toy industry, students can apply for summer or part-time jobs with local manufacturers. The most likely areas to find jobs are in assembly work, sales, and marketing. A large portion of toys sell in the period before Christmas, so toy companies must have their products ready ahead of time. The months from July through September are usually the busiest in the year, and jobs may be most available during this time. Students with younger siblings can do some personal research right at home; determine which types of toys your brother or sister prefers and why. It might also prove worthwhile to spend some time at neighborhood day care centers to observe the habits of young children with their toys.

Employers

The most popular locations for toy companies in the United States are the largest cities and states, such as New York, Los Angeles, Chicago, and San Francisco, as well as Washington, Texas, Florida, New Jersey, Connecticut, and Pennsylvania.

Some workers are employed on a temporary basis manufacturing toys during the busiest season—that is, before Christmas.

Starting Out

For entry-level positions in the toy industry, job seekers can contact the personnel offices of toy manufacturers. This is true for most factory jobs, whether applicants are looking for engineering, management, marketing, or factory production jobs. Some job listings and information may be available at the local offices of the state employment service, at local union offices, or in newspaper classified ads.

Advancement

In general, advancement to better jobs and higher pay depends on acquiring skills and gaining seniority. Some production workers advance by learning to operate more complex machinery. Reliable, experienced workers in production jobs might be promoted to supervisory positions. Professional and management staff can progress in various ways depending on their areas of expertise.

Earnings

Newly hired production workers may be paid at rates not much above the federal minimum wage. With experience, they may average as much as $9.50 an hour. Many workers are paid on a piecework basis—that is, according to the amount of work they complete or the number of items they assemble. Machine operators usually earn more than assemblers who work by hand. During the peak production season from July to September, factory workers may have to work long shifts, and they are paid overtime rates for the extra hours.

Management and engineers are often paid a straight salary. Salary ranges vary from company to company and especially from job to job. For example, research and development employees can start at about $25,000 per year; some may eventually work their way up to $150,000 or more annually. Salary levels for these workers depend on their job responsibilities, experience, seniority, and quality of work. According to the *Occupational Outlook Handbook,* industrial designers overall had beginning salaries of about $27,000 in 1996, senior designers earned $45,000 and executives earned up to $140,000 annually. Beginning engineers averaged salaries around $38,000, growing to $59,000 at mid-level and $99,000 at senior level.

Work Environment

The production floor of some toy factories is simply a large room in which workers perform routine tasks. A factory may employ as many as several hundred people to do production work. Some people work at machines, while others sit at tables or assembly lines. Some workers stand much of the day. Workers often have to meet production schedules and quotas, so they have to keep up a brisk work pace. Some people are bored by the repetition in many production jobs, because they must do the same few tasks over and over for long periods.

In smaller companies, the work may be highly seasonal. Getting the company's products ready for selling in the Christmas season, and to some extent, the Easter season, can mean that employees are asked to put in 10 or more hours of work a day. And if the company makes a product that becomes extremely popular, workers may have to scramble to make enough of the item to keep up with demand. But in the off-peak season, usually the winter months, and in average conditions, production workers may have reduced hours or they may be laid off. In many shops, some production workers are employed only five or six months a year. Management and other professional employees work year-round. They may need to put in overtime hours during peak seasons or before trade shows, but they do not earn overtime pay.

Outlook

According to the U.S. Industrial Outlook from the Department of Commerce, the toy industry experienced 3.4 percent growth in 1998. In the foreseeable future, employment in the U.S. toy industry probably will not undergo much net change. Sales of toys and games for both children and adults are increasing. On the other hand, some of these products, notably some electronic video games, are imports from abroad. If toy preferences change, employment patterns may shift in coming years but in ways that are hard to predict now.

There is a fairly high rate of job turnover among production workers, who make up a large part of the total workforce in the toy industry. Because of the low pay, repetitive work, and seasonal fluctuations in workloads, production workers may quit after a time to find more stable employment. Consequently, manufacturers are often looking to hire new production workers.

For More Information

The following labor union represents toy industry workers.

Amalgamated, Industrial and Toy and Novelty Workers of America
147 East 26th Street
New York, NY 10010
Tel: 212-889-8180

This trade association is for U.S. producers and importers of toys and holiday decorations.

Toy Manufacturers of America
200 Fifth Avenue, Room 740
New York, NY 10010
Tel: 212-675-1141
Email: info@toy-tma.org
Web: http://www.toy-tma.org/index.html

For information on a four-year program in toy design, contact:

Fashion Institute of Technology
227 W. 27th St.
New York, NY 10001
Tel: 212-217-7133
Web: http://www.toy-tma.org/ati/fit

Index